For Ted: —

Who comes belatedly to the polyamines.

But better late than never.

Seymour

Introduction

to the

POLYAMINES

Introduction to the POLYAMINES

Seymour S. Cohen

School of Medicine
University of Pennsylvania

Prentice-Hall, Inc., Englewood Cliffs, New Jersey

Current printing (last digit):

10 9 8 7 6 5 4 3 2 1

13-493114-9

Library of Congress Catalog Card No. 79-152449
Printed in the United States of America

PRENTICE-HALL INTERNATIONAL, INC., *London*
PRENTICE-HALL OF AUSTRALIA, PTY. LTD., *Sydney*
PRENTICE-HALL OF CANADA, LTD., *Toronto*
PRENTICE-HALL OF INDIA PRIVATE LIMITED, *New Delhi*
PRENTICE-HALL OF JAPAN, INC., *Tokyo*

Preface

This small book comprises four lectures prepared for presentation as a course at the Collège de France in May, 1970. A course at the Collège is intended to apprise the scholarly community of the recent developments at the frontiers of knowledge. To quote the tercentenary volume of the Collège:

> Il ressort que le Collège de France a servi souvent, selon l'esprit de son royal fondateur, à des enseignements nouveaux, qui n'avaient pas encore reçu ailleurs droit de cité. C'est ce qui a fait dire à Ernest Renan qu'il était spécialement destiné à la science, "en voie de se faire."

I decided that this charge to the faculty of the Collège could be met by a stimulating review of the literature on the physiology and biochemistry of the polyamines as well as by a discussion of some of our recent and as yet unpublished data in this area. It is my view that the scientific community is scarcely aware of the breadth, depth, and possible import of available information on these substances. Modern biochemistry has been concerned mainly with anions, including polyanions such as the nucleic acids and most proteins, and, for a variety of reasons, has tended to neglect the study of the cations. These lectures are intended to show that existing data on the organic cations should be more widely considered in approaches to cellular physiology. I believe that we are on the verge of new significant generalizations concerning the molecular roles of these compounds in the structure and function of cells and their components. This seems particularly true concerning the roles of the polyamines with respect to the biochemistry of the nucleic acids, for which polyamines serve as organic cations in many circumstances. The presentation of data relevant to this perspective of future research is thus the main intent of the lectures and of their publication for a wider audience.

I am pleased to have been invited to present my course by Professor

François Jacob, an old friend. It was entirely appropriate that these discussions on the compounds serving as cations for the nucleic acids should be sponsored at the Collège de France by a Laboratory of Genetics. Genetics is a discipline that has done much to help clarify the function and structure of the nucleic acids. It was also appropriate because the polyamine salt, spermine phosphate, was discovered independently in 1791 by one of the earliest Professors of Chemistry at the Collège, Nicolas-Louis Vauquelin, who was also one of the fathers of metallurgy and organic chemistry. The tradition of Vauquelin in the study of these natural products has been maintained at the Collège in the Laboratory led by Professor Jean Roche. With Thoai and Robin, this Laboratory has made significant contributions to our knowledge of the biochemistry of guanidino compounds, including the naturally occurring diamidino polyamines.

The lectures were given in French in a form somewhat more abbreviated than that herein presented. The English version had been translated into French by several very kind collaborators, including my daughter, Sara E. Cohen, and Dr. Yvonne Khouvine, to whom I am very grateful. The problem of determining whether the translation fitted my meaning and intent unquestionably made me think more carefully about what I was saying. I was also happy to have the attentive ear of my wife, who was as surprised and pleased as I was to have the entire complex operation come out as well as we think it did.

During our stay in Paris the hospitality of our numerous friends at the Pasteur Institute, Institut du Radium, Institut de Biologie Physico-Chimique, Orsay, Gif-sur-Yvette, etc. was an overwhelming experience that extended from lecture to lecture. Nevertheless, the course remained, around which we organized our thoughts and activities during this month. It is a pleasure to recall that the absence of other responsibilities permitted a greater care and freedom in the preparation of the lectures than I have enjoyed in many other scholarly activities. In the course of this preparation, it occurred to me that a scientific community greater than that represented by my audience at the Collège might also benefit from a broad but not detailed presentation of the data on the polyamines and indications of the scope of the biological phenomena to which they are applicable. These considerations have determined the content and form of the lectures and of this book, which makes no pretense of being comprehensive and thorough in describing a field that is "en voie de se faire."

SEYMOUR S. COHEN

Philadelphia, Pennsylvania

Contents

Introduction
to the
POLYAMINES

Chapter 1

In the Path of Leeuwenhoek, Vauquelin, Charcot, and Rosenheim

"All this has been said before—but since nobody listened, it must be said again."

ANDRÉ GIDE

The polyamines are low molecular aliphatic nonprotein nitrogenous bases. Putrescine and cadaverine are the simple diamines, 1,4-diaminobutane and 1,5-diaminopentane respectively. Spermidine and spermine are aminopropyl derivatives of putrescine. These structures are given below:

$H_2N-(CH_2)_4-NH_2$ **1,4-Diaminobutane** (putrescine)

$H_2N-(CH_2)_5-NH_2$ **1,5-Diaminopentane** (cadaverine)

$H_2N-(CH_2)_4-NH-(CH_2)_3-NH_2$ **Spermidine**

$H_2N-(CH_2)_3-NH-(CH_2)_4-NH-(CH_2)_3-NH_2$ **Spermine**

The history of the polyamines is some two hundred years older than that of the nucleic acids. The former began in 1678 with the discovery by Antony van Leeuwenhoek of the crystallization of spermine phosphate from human semen (223). Leeuwenhoek had begun his observations with his primitive microscope in 1674, and in three years he had advanced to the study of human sperm (88). Continuing observation of a single specimen of semen had permitted him to observe the gradual formation of colorless crystals. He reported this observation to the Royal Society in London in the famous letter (223) in which he first described spermatozoa. We know today that the slow deposition of spermine as the phosphate requires a prior liberation of inorganic phosphate via the enzymatic hydrolysis of the phosphorylcholine present in the seminal fluid. The salt, spermine phosphate, crystallizes easily and spontaneously.

Nicolas Vauquelin also reported the formation of the crystals in his single paper on human semen, published in 1791 (443). He was apparently unfamiliar with the prior report of Leeuwenhoek. His own paper was largely concerned with the first published observations and experiments on the clotting and liquefaction of semen, but he also records the slow crystallization of spermine phosphate, as well as its relative insolubility in water and ethanol.

The story of Vauquelin's discovery when he was a young man of 28 and the significance for reproductive physiology of the numerous observations in his paper are ably described by Williams-Ashman (455). In 1791 Vauquelin was the assistant and demonstrator for Antoine François de Fourcroy, Professor and Director of the Lycée of the Rue de Valois, in which Lavoisier also had participated. As Fourcroy's assistant, Vauquelin had discovered the combination of phosphorus with chlorine some years earlier. His subsequent professional career and somewhat curious personal arrangements were closely tied to Fourcroy (375). Vauquelin became the Professor of Chemistry at the Collège de France in 1801; he left this position to take the chair at the Musée d'Histoire Naturelle. After Fourcroy's death in 1809, he took over the chair of chemistry at the Faculty of Medicine. In that post he made many studies on urine and on calculi, studies that were of considerable significance in urology. By the time of his death in 1829, Vauquelin [1] was one of the great figures of French chemistry, renowned in addition for his discoveries of chromium, beryllium, and the first known amino acid, asparagine.

Many naturalists (physicians, pathologists, etc.) rediscovered the crys-

---

[1] Historians of chemistry have said (375a) that "his real devotion was to his work, and no chemist can have worked harder or with greater enthusiasm." Cuvier described him as "entirely a chemist, a chemist every day of his life, and throughout every day." We note that Vauquelin did his work through the periods of the Revolution, the Directory, Napoleon, and the Bourbon Restoration and many of his close colleagues, e.g., Fourcroy, were actively involved in the significant political movements of the time (375). Nevertheless, we find M. Vauquelin deeply immersed in the development of his discipline. He ranged broadly wherever this discipline could probe, and until his death turned up significant new observations and discoveries at every step. I am happy to recognize a man who did not lose his enthusiasm, his skill, and his creativity and who made his discipline the clearest instrument of self-expression. His discoveries, such as that of chromium and of amino acids, have had important application in the development of industry and in the improvement of human nutrition. He provides another example to a certain section of youth who, disturbed by the deterioration of the human condition, suggest in many "civilized" countries that scientists are wasting their time in games that cannot significantly affect the future of man on this planet, except for the worse.

tals that deposit in aged semen; however, the nature of this crystalline substance was not clarified for over a century (455). Charcot, the famous physician and teacher of Freud, published in 1853 (he was only 28 at the time) a paper on the crystallization of spermine phosphate. In 1865 the crystals were redescribed by Boettcher, who thought the substance was similar to proteins. In 1878 Schreiner showed that the crystals were a phosphate salt of an organic base, identical with similar crystals obtained from many tissues. Fuerbringer proved in 1881 that the prostate gland was the primary source of most of the spermine of semen. The name "spermine" was given to the base by Ladenburg and Abel in 1888, but the correct structure awaited the important work of Otto Rosenheim from 1924 to 1926. Rosenheim and his collaborators (329, 93) synthesized spermine and the related base, spermidine, which they had also discovered (94); their findings were confirmed by a group led by Wrede in 1927.

Isolation and Structure of Spermine

Rosenheim's contributions to this problem were numerous and rigorous.[2] Despite the early references to crystalline spermine phosphate as "Boettcher's crystals" and the elaboration of a method of isolation by Schreiner in 1878, the successful reisolation of this material awaited Rosenheim's adoption in 1924 of Vauquelin's original technique. This consisted of permitting the crystals to form for several hours in semen, diluting somewhat with a salt solution, and then centrifuging the precipitate. The sediment was washed with water to remove seminal proteins. The

[2] The papers of Sigmund Otto Rosenheim (1871–1955) on spermine and spermidine, their isolation, characterization, and synthesis, are among the clearest and most elegant expressions of the biochemical tradition in which I was trained. The publication of these papers immediately preceded Rosenheim's election to the Royal Society, but he continued to make discoveries and is probably most widely remembered for his later work revealing the structure of cholesterol (207).

At the Collège de France, in which molecular biology is so well represented, I have taken this opportunity to call the attention of my audience to the esthetic qualities of classical biochemistry. Rosenheim's terse papers contain, in the path of discovering new compounds existing in nature, the simple pleasures of crystallization, of handling and accounting for substance, and the arithmetic of stoichiometry in the analysis of structure. We also find the satisfactions of synthesis, which confirms our deductions concerning structure, reveals the powers of man and his science, and in furnishing substance according to our wish and need provides the jumping-off point for new investigations. It would be unfortunate if the training of molecular biologists did not include the essential skills of discovering, characterizing, and transforming substance, if only for the esthetic pleasures involved.

residue was recrystallized from hot water; crystalline spermine phosphate was obtained in amounts of about 10–30 mg per 10 ml of semen.

Spermine was also shown to be separable from semen by dialysis, by butanol extraction at high pH, or by steam distillation from alkaline solution. The crystallographic properties and melting points of the crystalline phosphate, picrolonate, chloroplatinate, and chloroaurate were identical with these characteristics of comparable materials isolated from the testis, the pancreas, and yeast.

The reactions of the compound with H_2O_2, $KMnO_4$, HNO_2, etc. and its lack of optical activity indicated that the compound is a saturated aliphatic base that does not contain any asymmetric carbon atoms. It formed a tetrabenzoyl derivative and, on exhaustive methylation, 10 methyl groups were taken up, indicating the presence of two primary amino groups and two secondary nitrogens. Degradation by heat led to the formation of pyrrolidine, a moiety similarly produced on distillation of tetramethylenediamine (putrescine) dihydrochloride. Degradation of decamethyl spermine sulfide led to the recovery of tetramethyl trimethylene diamine.

The pure base did not possess the characteristic odor that is usually associated with semen. A volatile base with this odor was produced by mild oxidation and was thought to be N-γ-aminopropyl pyrroline.

Spermine was the first natural compound found to contain propanediamine moieties. However, in 1887 Brieger had isolated three bases from cultures of *V. cholerae,* i.e., putrescine (1,4-diaminobutane), cadaverine (1,5-diaminopentane), and a third whose elementary composition was consistent with 1,3-diaminopropane. This base has been reported to exist in many animal tissues (and human semen), as well as in many bacterial cultures.

Rosenheim pointed out that spermine is not detectable in washed sperm and is absent from bull semen. The bull appears to lack a functional prostate and, in general, the spermine content of semen relates to the activity of this gland. The content of the two bases, spermine and choline, is highest in human semen, and the detection of the insoluble spermine picrate and flavianate are two tests for human semen in forensic medicine.

The compound does not activate spermatozoa of various species; therefore, it most likely is inessential to sperm motility or survival. What then is the function of the compound in semen? It is known that spermine at high concentrations is sufficiently bacteriostatic to prevent the growth of staphylococci in human semen. On the other hand, bull semen is notoriously infected with microorganisms (gonococci for example), and many observers think that the synthesis of RNA and protein detected in preparations of bull sperm is caused by microbes present in the fluid. Furthermore, spermine oxidase in human seminal fluid oxidizes spermine to a compound that is toxic at low concentrations to many bacteria and viruses. In 1952 Hirsch and Dubos described the isolation from many tissues of a compound highly inhibitory to tubercle bacillus (166). This compound was then shown to be spermine. Its toxicity to mycobacteria depended on its conversion to an active inhibitor by a serum amine oxidase (164, 165). Other possible roles of the polyamines in seminal fluids have been reviewed recently by Williams-Ashman and Lockwood (455a).

Spermidine and the Synthesis of Triamines

Rosenheim and his colleagues reported in 1924 that a soluble phosphate accompanied spermine in the isolation procedure (93). This organic base was named spermidine. Three years later, the isolated compound was identified as a salt of spermidine. Its structure was shown to be γ-aminopropyl amino butylamine and was confirmed by synthesis by the same workers (94). Intermediates involved in the synthesis were obtained from the sequence of reactions (Fig. 1) with putrescine and phenoxypropyl bromide leading to spermine. As might be expected, even when equimolecular amounts of the reagents are used, some spermine is produced in this reaction sequence. It is appropriate to note that many of the samples of spermidine obtained from "reliable" commercial sources come as grossly contaminated products, containing as much as 20–60% spermine.

Possible structures

(a) $NH_2CH_2CH_2CH_2NHCH_2CH_2CH_2CH_2NHCH_2CH_2CH_2NH_2$

(b) $NH_2CH_2CH_2CH_2CH_2NHCH_2CH_2CH_2NHCH_2CH_2CH_2NH_2$

Synthesis (94)

$$H_2NCH_2CH_2CH_2CH_2NH_2 \xrightarrow[\text{Excess}]{\phi OCH_2CH_2CH_2Br}$$

$$\phi OCH_2CH_2CH_2NH\ CH_2CH_2CH_2CH_2NHCH_2CH_2CH_2O\phi$$

$$\downarrow HBr$$

$$BrCH_2CH_2CH_2NHCH_2CH_2CH_2CH_2\ NHCH_2CH_2CH_2Br$$

$$\downarrow NH_3$$

Structure (a) = Spermine

= α, δ-*bis* [γ-aminopropylamino] butane

Fig. 1. The synthesis of spermine and proof of structure (94).

More reliable syntheses of spermidine have been devised, as in the reactions of Danzig and Schultz (76) or of Jackson (184) presented in Fig. 2. Minor variations of these have been employed in the synthesis of analogs of both spermidine and spermine. For example, good yields of cyanoethylated derivatives are formed by dropping one or two equivalents of acrylonitrile into liquid diamine (183). A sponge nickel catalyst facilitates reduction by H_2 at moderate pressure in ammoniacal ethanol. The products have been isolated and purified by vacuum distillation.

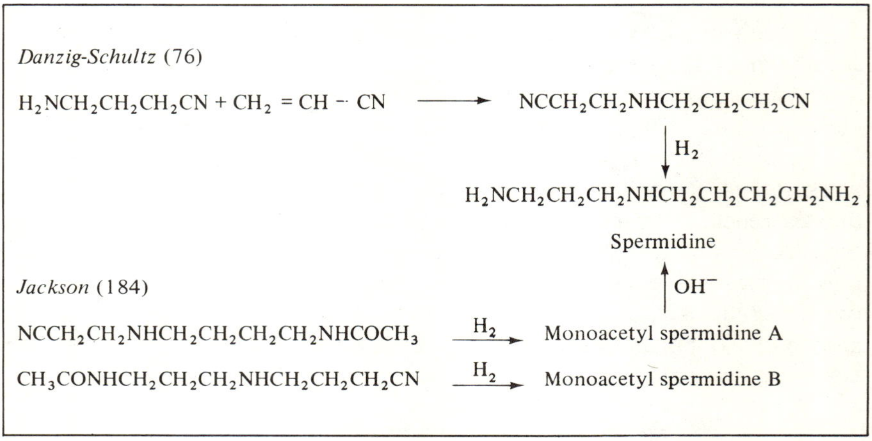

Fig. 2. Synthesis of spermidine and its monoacetyl derivatives (76, 184).

Analogs (183) in which the butyl moiety has been replaced by C_2 to C_{12} chains are as inhibitory to KB cells (human epidermoid carcinoma) in calf serum as are spermidine and spermine themselves, i.e., 50% inhibition at about 10^{-5} M. As a result, observers have realized that the inhibitory qualities of the polyamine analogs toward cellular and bacterial test systems arise largely from the presence of the enzyme, diamine oxidase, in the serum. Nevertheless, some of these analogs do produce significant inhibitions of mouse tumors and leukemias.

Oxidation of Polyamines

How shall we explore the functions of the polyamines? Are they merely potential antibiotics? If presumed intracellular roles are to be explored, we must have systems that will not convert exogenously supplied amines to toxic products. Enzymes from a variety of sources oxidize the polyamines, and it is useful to know what to avoid.

In 1961 it was observed that the natural polyamines, spermine and spermidine, were toxic to a number of mammalian cells, including Ehrlich ascites cells in tissue culture. Israeli workers (Halevy et al.), who had explored the requirement of polyamines in microbial nutrition, similarly observed that spermine at 2.5×10^{-6} M exerts a toxic and rapidly destructive effect on chick embryo fibroblasts cultured in a medium fortified with 2% calf serum (144). Spermidine was half as active, whereas putrescine was not toxic at a 20-fold greater concentration. Even as described by Hirsch for the inhibition of mycobacterial growth, these effects were due to the action on the polyamines of an amine oxidase present in calf serum. Spermidine, but not putrescine, is oxidized by this enzyme. It appears that the enzyme is present in the sera of all ruminants as well as in that of some non-ruminants, e.g., the hippopotamus, whose digestive systems contain accessory caeca in which some unusual fermentations are harbored (32). The sera of most non-ruminants are low in amine oxidases or contain oxidases that are inactive on the polyamines. Spermine is thus non-toxic below 10^{-4} M when offered to tissue cultures containing human sera.

The spermine oxidase of beef plasma is a copper enzyme inhibited by cyanide, and also by carbonyl reagents, such as semicarbazide. This protein, which has been crystallized, is representative of a group of copper-containing amine oxidases, e.g., pea-seedling amine oxidase and diamine oxidase of pig kidney, which are sensitive to cyanide and carbonyl reagents. The beef spermine oxidase has a molecular weight of 170,000 and tends to aggregate (2); it is dissociated by 8 M guanidine to two equivalent polypeptide chains of 87,000 mol wt.

The serum enzyme also oxidizes benzylamine (and some other amines) to H_2O_2 and benzaldehyde, which is detectable by its ultraviolet absorption at 250 mμ. Unlike many other amine oxidases, the enzyme does not oxidize tryptamine, serotonin, histamine, putrescine, cadaverine, etc. This specificity has served to distinguish this enzyme from the structure-bound mitochondrial monoamine oxidase, which does oxidize such substrates as serotonin and is inert on spermine. Surprisingly, Gorkin and Tatyanenko (135) have reported a method of converting the mitochondrial enzyme to a form that is capable of oxidizing spermine, as well as other substrates of the soluble Cu-containing amine oxidase. However, the relation of the mitochondrial enzyme to the plasma enzyme is still unknown.

The oxidation products of spermine and spermidine by the purified enzyme have been described by the Tabors and Bachrach (413) as aminoaldehydes; the inhibitory properties of the aminoaldehydes have been explored largely by Bachrach. It can be seen in Fig. 3 that spermine gives rise to a dialdehyde with liberation of two moles of NH_3 from the two aminopropyl sidechains. The dialdehyde was characterized after reduction to the N,N'-*bis* (3-hydroxypropyl)-1,4-diaminobutane. On the other hand, spermidine is oxidized at the single aminopropyl end to the monoalde-

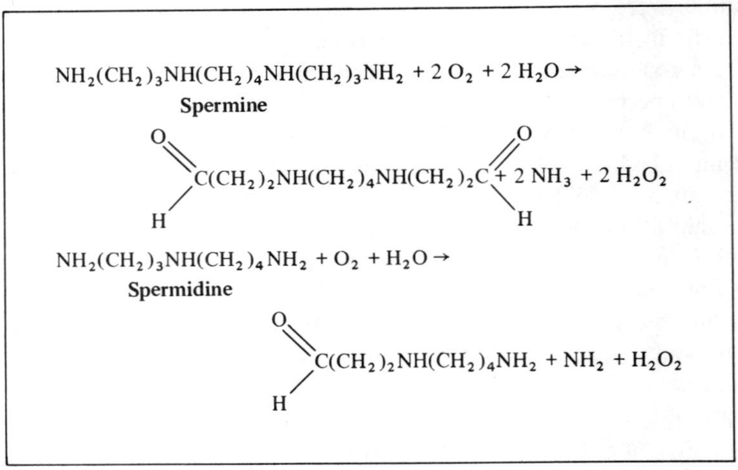

Fig. 3. The enzymatic oxidation of spermine and spermidine (413).

hyde. It was subsequently shown that monoacetylspermidine A, possessing the acetyl at the aminobutyl end, can be oxidized by the enzyme, whereas monoacetylspermidine B, acetylated at the aminopropyl end, was inert.

These aldehydes are much less stable than the parent compounds; the spermidine product will decompose non-enzymatically to putrescine whereas the intermediary monoaldehyde derived from spermine can degrade spontaneously to spermidine. It has been found by Alarcon that acrolein is also produced in some as yet unknown reaction sequence during the oxidation of spermine and spermidine by a plasma amine oxidase (5). Acrolein has been shown to be quite toxic to cells in its own right; therefore, Alarcon has asked if this substance is not the true inhibitor produced non-enzymatically following enzymatic oxidation of spermine and spermidine.

Inhibitory Activities of Oxidized Polyamines

The unstable dialdehydes generated by the purified oxidase are commercially available as acetals, from which the aldehydes are readily regenerated. In recent studies, Bachrach and his colleagues have shown that these purified compounds inhibit the growth of cells in tissue culture producing a relatively rapid inhibition of protein and nucleic acid synthesis. Oxidized spermine inhibits *E. coli* at about 10^{-3} M and rapidly prevents the induction of β-galactosidase (19), as shown in Fig. 4. At this concentration, the dialdehyde clearly prevents RNA synthesis before protein synthesis. The substance blocks the synthesis of m-RNA by serving as an

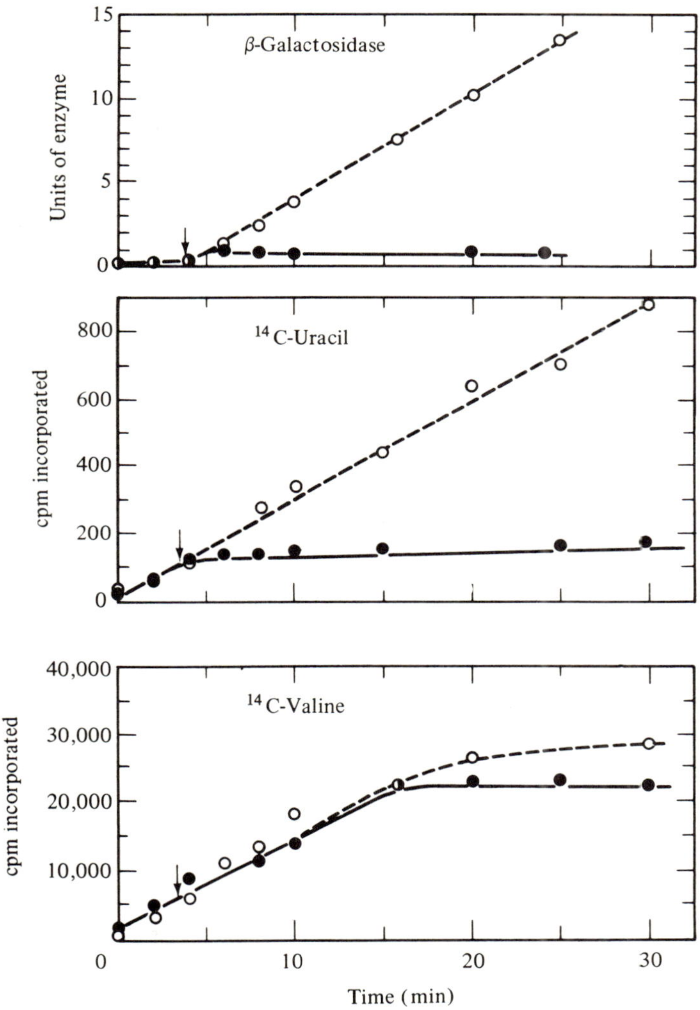

Fig. 4. The action of oxidized spermine on *E. coli* (19). The cells were induced by β-D-thiomethyl galactoside in the presence of isotopic uracil or valine. Oxidized spermine (120 μg/ml) was added to one of the two cultures at the arrow. The appearance of enzyme and the incorporation of the precursors were measured.

agent acting on intact Gram-negative bacteria in a manner similar to the action of actinomycin D on Gram-positive organisms. Indeed, the dialdehyde has been used to inhibit RNA synthesis in *E. coli* in order to determine the half-life of various messenger RNAs.

The compound inhibits various RNA viruses, like tobacco mosaic virus (22) and MS2 (24), as well as DNA viruses, including many phages such as

T_3, T_5, T_7, ϕ80, λ, etc. (24, 285). T-even phages, as well as ϕX174, P_{22}, SP_{10}, were not readily inactivated (24, 285). The data on T_5 inactivated by the dialdehyde indicated that inactive DNA, containing tritium originally present in the oxidized spermine, was injected from the inactive virus into the bacterial cells (16). This DNA is not replicated nor does it serve as a template in synthesis of complementary RNA (17).

This observation led Bachrach to study the action of oxidized spermine on isolated DNA. Two types of linkage were detected (13), one by reversible electrostatic bonds effected via the secondary amino groups and the other via irreversible covalent bonds effected through the carbonyl groups. The latter reaction produces cross-links between the strands, resulting in marked increases of melting temperature (T_m) and particularly in reduction of the maximal hyperchromicity, as shown in Fig. 5. The latter result is unlike that with oxidized spermidine, which increases T_m but cannot reduce the hyperchromicity produced by denaturation. The binding by carbonyl groups is observed to be maximal with guanine nucleotides (97) and G-C-

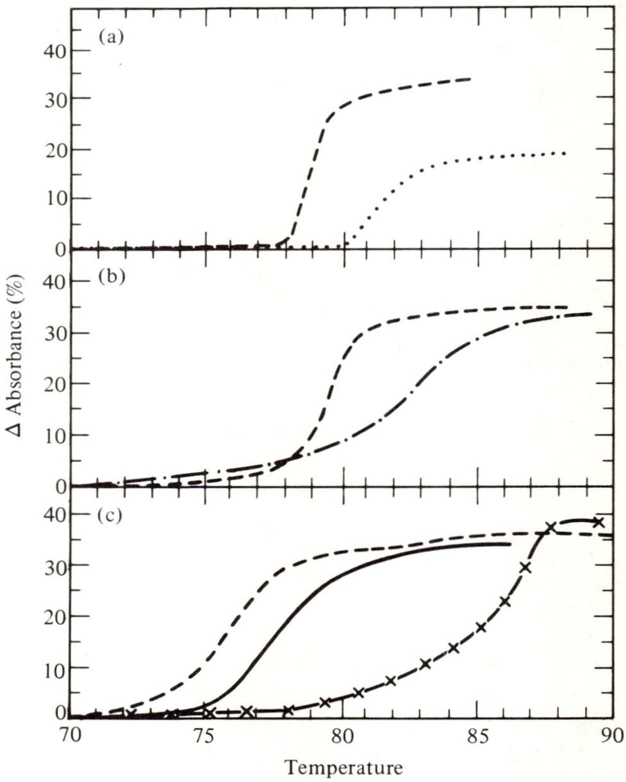

Fig. 5. Absorbance-temperature profiles (13). *E. coli* DNA was incubated with (a) oxidized spermine (dotted curve); (b) oxidized spermidine (dot-dash curve); (c) spermine (x-x-x-x), spermidine (solid curve). DNA without addition, control (dashed curve).

rich DNA (14). Such binding is eliminated by reduction of the aldehyde with sodium borohydride or by deamination of the nucleic acids (13). Native bihelical DNA binds up to 15 molecules of oxidized spermine per 10^3 nucleotide units. The binding of the dialdehyde to the DNA primer inhibits the in vitro syntheses effected by both DNA polymerase and the DNA-dependent RNA polymerase (20). It can be noted [Fig. 5(c)] that the unoxidized spermine is more effective than the oxidized compound in increasing the melting temperatures of DNA.

The reversible binding via the secondary nitrogens of the dialdehyde suggests that the central putrescine moiety may play a significant role in positioning the inhibitory carbonyl groups for its irreversible binding. This positioning would then be comparable to the positioning postulated for the comparable moiety in chloroquine and other intercalating molecules whose reactions with the bases in the central helical structure is facilitated by the initial electrostatic binding of the drug to the external phosphates bounding the groove in which the aromatic ring of the drug is inserted. This structure is described later in greater detail in connection with the model of polyamine binding to DNA proposed by Tsuboi (438) and Liquori (231).

On the other hand, it may be asked if DNA treated with the dialdehyde derived from spermine or the aminoaldehyde derived from spermidine does in fact contain the putrescine moiety of these compounds. If acrolein is truly the inhibitory agent, released from the unstable aldehydes, as postulated by Alarcon, the relative loss of putrescine carbon and conservation of aminopropyl carbon should be easily demonstrable. Unfortunately, such critical experiments have not yet been performed to substantiate or to challenge the molecular mechanism (which involves the entire compound) suggested by Bachrach. In some experiments on the toxicity of oxidized spermine, labeled with ^{14}C exclusively in the diaminobutane moiety, the isotope was incorporated into the cells (11). Unfortunately, these experiments do not appear to answer rigorously the specific question of the retention of both the diaminobutane and aminopropyl moieties.

Finally, we note that both spermine and oxidized spermine can protect DNA against various nucleases (15). Thus, the nucleate to which polyamine is attached has been made refractory to the enzyme. Data of this type with deoxyribonuclease I are presented in Fig. 6.

Some Enzymatic and Chemical Assays of Spermidine and Spermine

Simple fluorescent or isotopic assays have been developed for diamine oxidase or monoamine oxidase. The relatively specific oxidation of spermidine and spermine by the serum amine oxidase to a fluorescent product in the presence of resorcinol was used initially to permit the estimation of these components at the level of 0.2–1 μmoles or that amount contained in

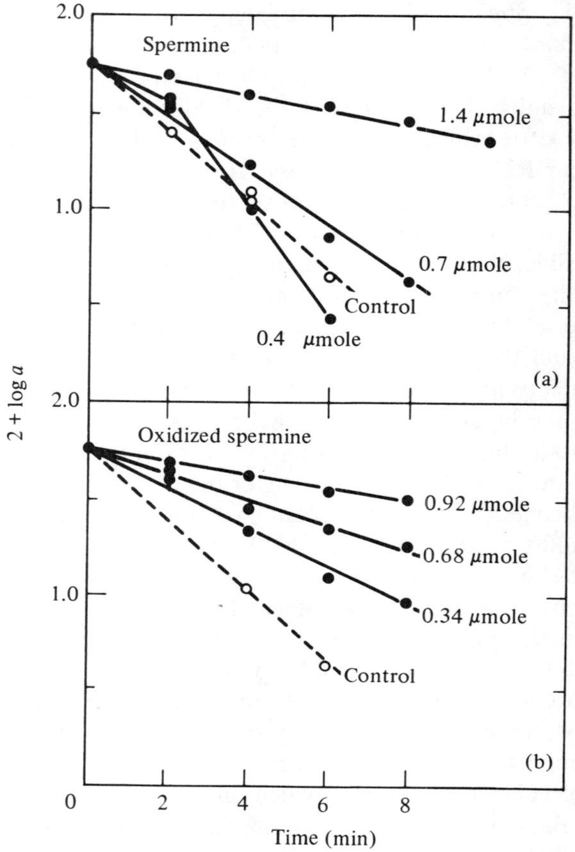

Fig. 6. The effect of polyamines on the degradation of DNA by deoxyribonuclease I (15): (a) spermine; (b) oxidized spermine. Ordinate *a* stands for unhydrolyzed DNA (mg).

about a gram of rat liver (442, 441). This fluorescent substance in fact arises from the production of acrolein during the oxidation of the polyamine. As noted earlier, putrescine, cadaverine, etc. do not interfere with these analyses. An adaptation (23) of the assay of the aldehyde generated by the enzymatic oxidation has permitted a 10-fold increase in sensitivity, by coupling the aldehyde formed to N-methyl-2-benzothiazolone hydrazone hydrochloride, as in the reaction given below:

Reaction II

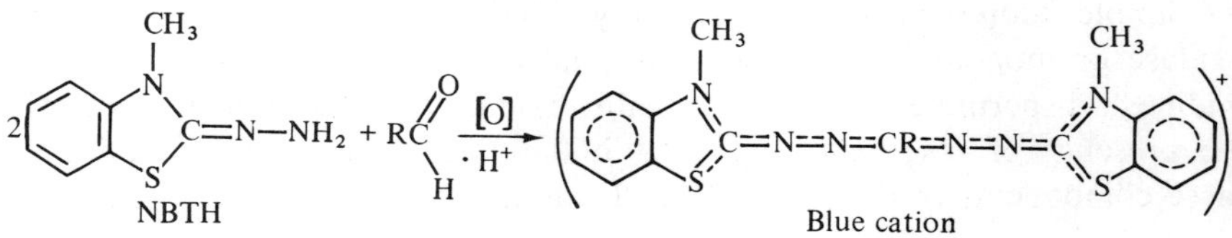

To estimate the polyamine components separately, we add an acid extract to an exchange resin and selectively elute the compounds, as described by Tabor, Rosenthal, and Tabor (420). Of course, such eluates, in the patterns presented in Fig. 7 or in other separation procedures sum-

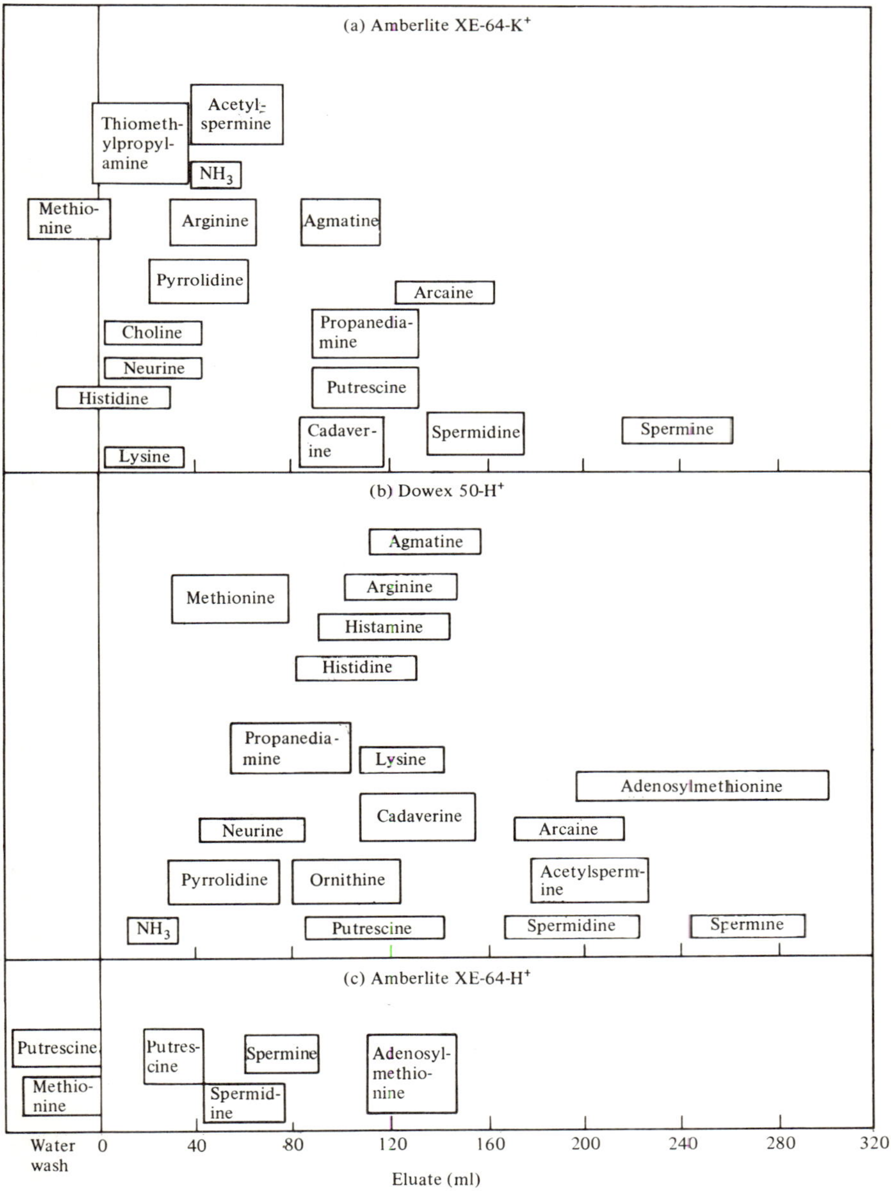

Fig. 7. Separation of amines on ion exchange columns (420).

marized elsewhere (425, 146), may also be analyzed by other procedures, using reactions such as those with dinitrofluorobenzene (332, 91), ninhydrin (146), or amido black (303). Some microfluorometric methods have been developed involving a direct reaction of polyamine with reagent. In the Kremzner method (212), spermidine and spermine form a fluorescent product with *o*-phthalaldehyde; this reaction can be used to estimate 1–10 mμmoles of these polyamines. By these methods it was found that rat blood contains about 1 μg of spermidine and 1.5 μg of spermine per ml, whereas rat liver may contain about 40 μg per g. Modifications of the Kremzner procedures have been even more sensitive (98, 177).

Very recently, my laboratory has turned to a new, relatively rapid method of analysis of polyamines. We can estimate 10^{-10} to 10^{-11} moles of these substances (353, 68, 87, 150). Fluorescent dansyl derivatives are formed by treatment of acid extracts with dimethylaminonaphthyl sulfonyl chloride in the presence of an excess of sodium carbonate. The derivatives are extracted in benzene and are readily separated as discrete spots in thin-layer chromatography. The fluorescent spots can be scanned in a fluorimeter to permit quantitation, as in Fig. 8. The amounts of the amines are proportional to the areas under the curves. With this method we can estimate the polyamines present in 0.2 ml of an *E. coli* culture at 10^8 organisms per ml or that present in nucleic acid samples of an absorbance of 0.01 at 260 mμ.

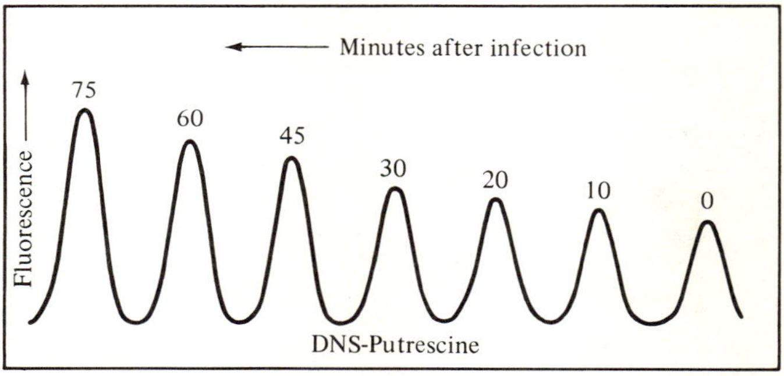

Fig. 8. Tracings of scannings of fluorimetric spots on a thin layer plate. The dansylated polyamines derived from aliquots of a culture of T4r⁺-infected *E. coli* strain B have been applied to thin layer plates containing silica and separated in cyclohexane-ethyl acetate (30:60). The tracings of the dansyl putrescine spots are presented. (Courtesy of Dr. A. Dion)

Some Bacterial Amine Oxidases

Assays have also been developed for various monoamine oxidases and diamine oxidases from pig kidney, rat organs, plant tissues, etc. that oxidize putrescine, cadaverine, histamine, tryptamine, and other amines. A radioactive substrate may be used to permit measurement of the formation of a

Serratia (53, 57)
spermidine $\longrightarrow$ H_2O_2 + δ-aminobutyraldehyde + 1,3-diaminopropane + Δ^1-pyrroline

Neisseria (54)
spermidine
spermine $\longrightarrow$ H_2O_2 + δ-aminobutyraldehyde + 1,3-diaminopropane

Pseudomonas (55)
spermidine $\longrightarrow$ H_2O_2 + putrescine + 3-aminoproprionaldehyde + β-alanine

Mycobacteria (56)
spermine $\longrightarrow$ H_2O_2 + spermidine + δ-aminobutyric acid + 1,3-diaminopropane + β-alanine

Micrococcus (58)
putrescine $\longrightarrow$ H_2O_2 + NH_3 + Δ^1-pyrroline

Fig. 9. Some bacterial oxidations of polyamines.

radioactive product (286). In a more general method, the hydrogen peroxide released is coupled to the formation of a fluorophore with homovanillic acid in the presence of horseradish peroxidase (387).

The oxidation of polyamines by various microorganisms has been known for many years. As shown in Fig. 9, the products of these reactions differ from those produced by mammalian enzymes. For example, an enzyme preparation from *Serratia marcescens* specifically converts spermidine to Δ^1-pyrroline and 1,3-diaminopropane. The pyrroline can be estimated colorimetrically in the range of 0.05–0.4 μmoles (18). *Neisseria* will oxidize both spermine and spermidine to pyrroline and diaminopropane (447).

A somewhat different oxidation has been observed with a *Pseudomonad* (288). In this reaction, spermidine is oxidized to putrescine and 3-aminoproprionaldehyde + H_2O_2. The aminoproprionaldehyde may be oxidized further to β-alanine.

The products of oxidation of spermine by *Mycobacterium smegmatis* have been reported to include spermidine, 1,3-diaminopropane, γ-aminobutyric acid, and β-alanine (21). No one has yet clarified the precise reaction sequence or sequences in this system.

Spermidine oxidation by the *Serratia* enzyme requires FAD and additional electron carriers (49), of which heme may be the major component (411). The *Serratia* enzyme is a dehydrogenase, therefore, rather than an oxidase. A similar FAD enzyme from *Micrococcus* is a putrescine oxidase (461) and is capable of converting putrescine to pyrroline, NH_3, and H_2O_2.

It has been known for many years that oxidation by the appendix of the spadix of arum lily species yields the highest known rate of oxygen consumption of any plant. This rate is sufficient to produce a significant warming of the tissue by 10–15°C over the temperature of the air. It appears that much of this oxidation, which is mainly cyanide-insensitive, is due to a light-stimulated action of the relatively non-specific plant diamine oxidase on amines in the plant. The plant is very rich in free amino acids, and over a dozen free amines have been detected, including compounds such as cadaverine and putrescine (376). The increased temperature produced by oxidation volatilizes these amines, which then attract insects active in cross-pollination. The size to which one of these lilies can grow with emergent appendix can be seen in Fig. 10.

Fig. 10. The Sumatra arum, *Amorphophallus titanum*, grown in a Netherlands greenhouse, with Hugo de Vries (at right) (254). Courtesy of Dr. C. G. G. J. van Steenis.

Some Other Polyamines and Their Distribution

The discovery of putrescine and cadaverine is attributed to the work of Brieger in 1885. He isolated these bases as heavy metal double salts from animal tissues. In the following year the compounds were shown to be identical with 1,4-diaminobutane and 1,5-diaminopentane, respectively, which were synthesized by Ladenburg in 1886. The bases were subsequently isolated from many bacterial cultures as well as from animal and plant tissues in various states, i.e., freshly harvested or under conditions of fermentation or putrefaction.

It is known that putrescine is derived from the decarboxylation of ornithine and cadaverine from the decarboxylation of lysine. No other route is known for the origin of cadaverine, nor has this reaction been described in plant or animal tissues. In the rat, subcutaneously administered cadaverine is oxidized and excreted as δ-amino valerianic acid, as presented in Fig. 11.

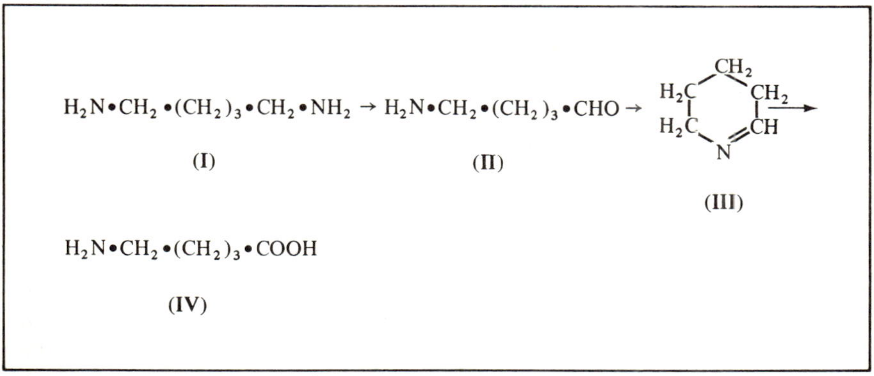

Fig. 11. The oxidation in the rat of cadaverine (I) to Δ^1-piperideine (III) and to δ-amino valerianic acid (82).

Numerous other reactions, shown in Fig. 12, lead to the formation of putrescine. Arginine is decarboxylated by arginine decarboxylase to agmatine. Agmatine can be hydrolyzed directly to putrescine and urea, or it may first form an intermediate, N-carbamylputrescine, which is hydrolyzed by an aminohydrolase to putrescine. The latter route is known in both plants and some bacteria. As we shall see later, the production of putrescine is under interesting physiological controls. For example, amine production is markedly increased in both plants and bacteria under conditions of K^+ deficiency or of acidity of the surrounding medium, suggestive of the existence of a homeostatic mechanism of proton elimination. The lysine

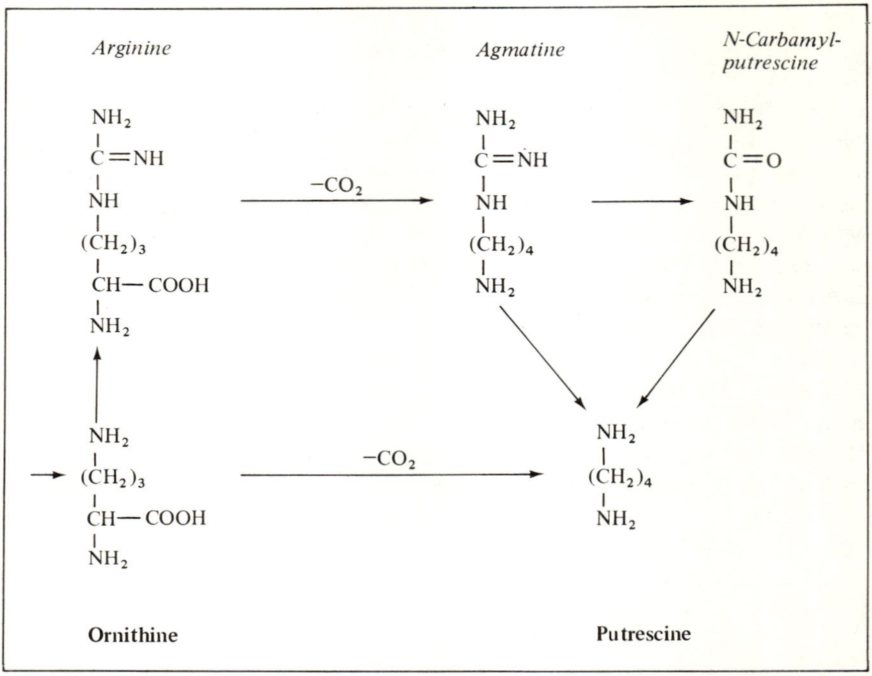

Fig. 12. Routes leading to the formation of putrescine.

decarboxylases of various bacteria have pH optima of 5 or less, and the appearance of cadaverine occurs most readily under fermentative conditions as the pH of the culture falls.

A number of other putrescine derivatives have been described (421). Feruloylputrescine (N-(4-aminobutyl)-4-hydroxy-3-methoxycinnamanide) is relatively high in all grapefruits and oranges, although absent from tangerines and lemons (452). Tetramethyl putrescine has been reported in higher plants (*Hyoscyamus* species), and python muscle is stated to contain monomethylputrescine (140). Japanese workers have recently detected the basic amino acid, putreanine or N-(4-aminobutyl)-3-aminoproprionic acid (Fig. 13) in brain but not in other organs (199). The role of this compound is unknown; it is thought to be the product of oxidation by brain of the aminoaldehyde generated from spermidine by an amine oxidase.

Several alkaloids have been found to be built of polyamines. For example, molecules such as pithecolobine and palustrin contain substituted diaminobutane moieties, whereas lunarine contains a diacyl spermidine that accounts for all the nitrogen of the molecule (421). The antibiotic peptide, edeine A, derived from *Bacillus brevis,* is terminated by a spermidine residue (199).

Diamidinoputrescine (arcaine) and diamidinospermidine (hirudonine)

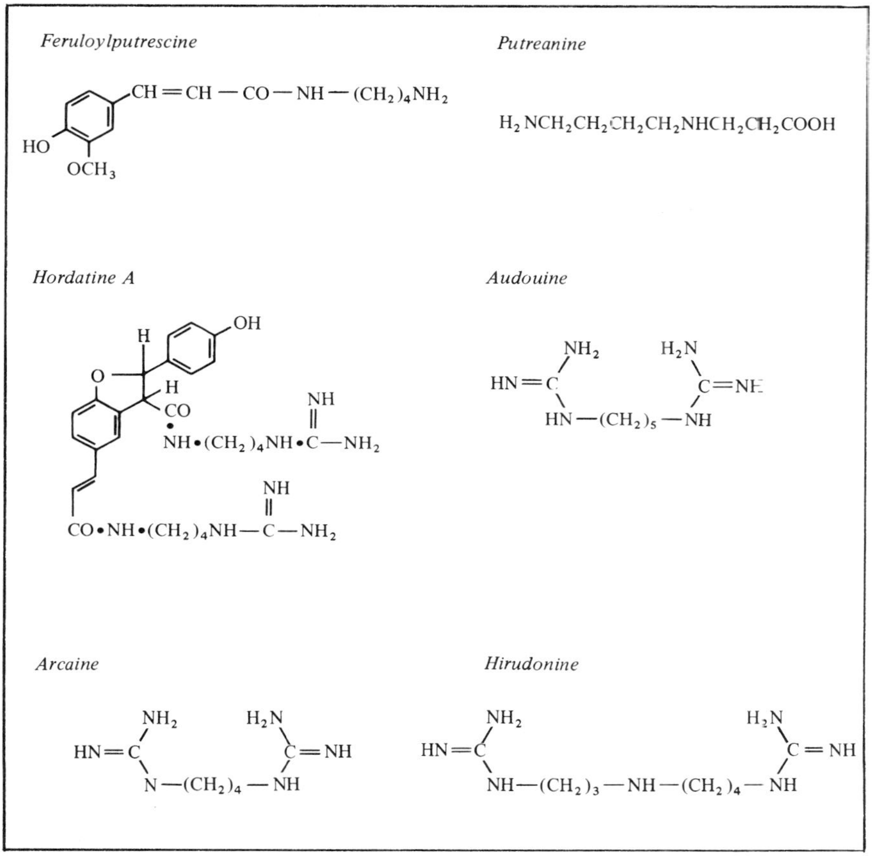

Fig. 13. Some naturally occurring polyamines.

have been found in several organisms (322), as described in the Laboratoire de Biochimie Générale Comparée of the Collège de France. Both guanido compounds are found in the leech, *Hirudo medicinalis*. Hirudonine has been synthesized by Robin and van Thoai (323), using a method that works very well. Robin and her collaborators (10, 321) describe how the amidine group in the leech is transferred stepwise from arginine to the corresponding polyamines, as in Fig. 14. Monoamidinated intermediates have been found; in fact, the antifungal agent derived from barley, hordatin A, is a disubstituted derivative of monoamidino putrescine (444). It may be mentioned that arcaine is a reasonable inhibitor of agmatine iminohydrolase (50% at millimolar concentrations).

A diamidinocadaverine (audouine) has been isolated from the marine polychaete annelid, *Audouinia tentaculata,* and synthesized in proof of structure (324). In this instance, we may ask if the cadaverine is derived from

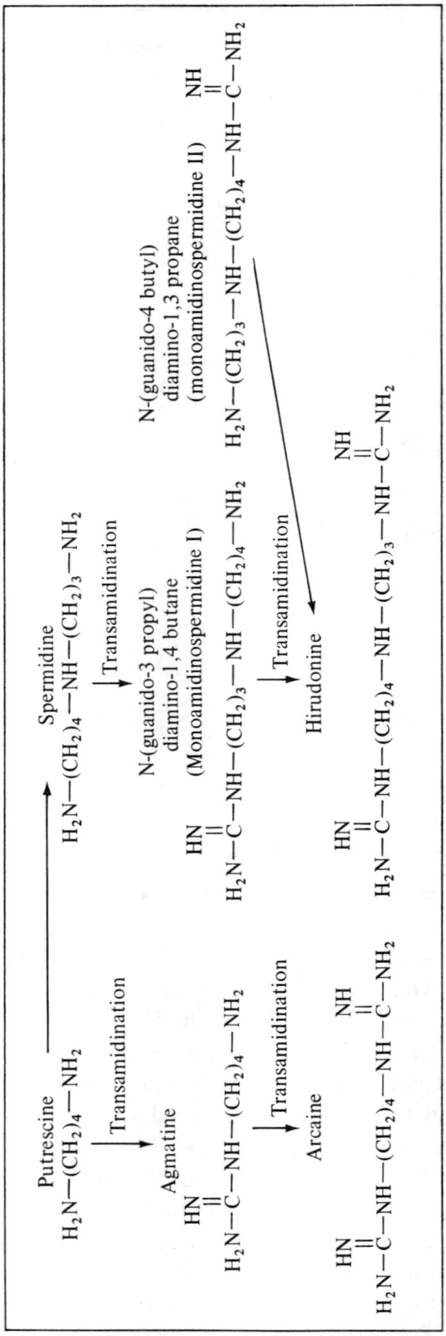

Fig. 14. The biosynthesis of arcaine and hirudonine in the leech, *Hirudo medicinalis* L.(10, 321).

lysine via an animal lysine decarboxylase or arises from a bacterial symbiont or parasite present in the annelid.

The subject of guanido derivatives is rather broad and I do not have the time to consider this matter further. However, I call your attention to the Ciba Foundation volume in honor of Professor Roche on the "Comparative Biochemistry of Arginine and Derivatives." Since it is well known that many basic guanido compounds, including various antibiotics, are very useful chemotherapeutic agents, you may wonder if these compounds prevent some functions of normal polyamines.

Two isomeric forms of monoacetylspermidine, in which either the aminobutyl or aminopropyl moieties are acetylated, have been isolated from cultures of *E. coli* (92). Monoacetyl spermidine B (acetyl on the aminopropyl moiety) has been isolated from normal human urine (278). Acetyl derivatives of putrescine and spermine (92) have also been isolated from certain types of bacterial cultures, as is discussed in Chapter 3 in some detail. Monoacetyl derivatives of putrescine and cadaverine have also been recovered from urine (295).

At this time we shall eliminate from future consideration a compound that was reported to exist uniquely in turnip yellow mosaic virus in significant amount; it clearly did not exist in the uninfected plant host. On the basis of gas-liquid chromatography of an extract of the virus, it was stated that the substance was the *bis* (3-aminopropyl) amine (197). The implication that virus infection could direct the synthesis of a new compound, for which some precedents do exist (63), had led to some interest in this report. However, the presence of this compound could not be confirmed recently by Beer and Kosuge (29). These workers have actually isolated a crystalline spermidine derivative from the virus. The polyamines present, i.e., spermidine and a small amount of spermine, amounted to 1% of the virus weight and could neutralize 20% of the virus nucleic acid. In addition, spermidine has been detected in several plant RNA viruses and in several DNA phages. No one has yet developed, however, a reproducible gas-liquid chromatography suitable for quantitation of free polyamines.

Polyamines in Gram-Positive Organisms

Despite the broad distribution of the organic cations, the latter have aroused only minimal interest among scientists. First, until the recent introduction of the dansyl method, estimating the amount of these substances in cells and salt-containing media has been slow, tedious, and difficult, despite the elegant chromatographic methods developed by the Tabors and the electrophoretic procedures of Raina. Second, the molecules are small and do not appear to have the capacity for specificity in combining with nucleic acids, a specificity that would whet scientific appetites. Third, the statement has appeared in the literature and has even been repeated in recent reviews that in general Gram-positive organisms contain little or none of the polyamines. Obviously, we would all prefer to work on compounds of general import in cell physiology.[3]

When the dansyl method of analysis is applied to a Gram-positive organism such as *B. subtilis,* we find that spermidine is present at high concentration and is easily estimated. Spermidine markedly increases during lag phase but then falls during the period of exponential growth to a level in which spermidine content per turbidity unit remains constant. In *Lactobacillus casei,* an organism reported to be very low in polyamine (154), it appears that the organism makes an enormous amount of spermidine in the lag phase (1400 μg per g dry weight) and then grows until that spermidine is diluted to a very low level (4.7 μg per g) (99). If the organism is harvested after 10 hr in stationary phase, spermidine is observed to fall still further to 1 μg per g, and the bacterium indeed appears to lack this polyamine. However, as is described later, polyamines markedly stimulate the growth of *L. casei* (204).

It appears that there is no unequivocal demonstration of a bacterium or of any other cell that can grow and multiply without polyamines. On the contrary, the ubiquitous distribution of the compounds appears well established. An unequivocal demonstration of cellular growth and multiplication in the absence of a natural polyamine or conceivably of a polyamine analog would raise serious questions in the mind of at least this observer.

The Biosynthesis of Spermidine

Having commented on the apparent ubiquity of these compounds, and having indicated the complexity of the pathways leading to putrescine, I shall at this point briefly outline our knowledge of the biosynthesis of spermidine and spermine. Procaryotic cells grown in a synthetic medium lacking spermine do not synthesize this polyamine but do contain putrescine and spermidine. On the other hand, the three polyamines are synthesized by all known eucaryotic cells capable of RNA synthesis.

We have recently examined the procaryotic blue-green alga *Plectonema boryanum,* provided by Dr. Moshe Shilo of the Hebrew University, and can affirm the apparent absence of spermine and the presence of spermidine and

[3] Recently it has been reported that a *Pseudomonas* strain contains putrescine but lacks spermidine. It has been stated, however, that in this organism, spermidine is replaced by 2-hydroxyputrescine (328). The latter compound is believed to stimulate the role of spermidine in its presence on ribosomes in a concentration inversely proportional to the external concentration of Mg^{++} (328). We shall explore these cellular and functional relations of the polyamines in subsequent chapters.

putrescine in this organism. Studies are being actively made on the capability of DNA phages to multiply in this blue-green alga and it may be asked if these phages also contain polyamines. On the other hand, the eucaryotic marine phytoflagellate, *Prymnesium parvum,* also provided by Dr. Shilo, does synthesize significant amounts of spermine, spermidine, and putrescine. Some interesting properties of a toxin elaborated by the latter organism are discussed later.

Spermine is an instance of a compound characteristic of eucaryotic cells. However, unlike some other substances of this type, its biosynthesis is not obviously tied to an evolutionary event consequent upon the evolution of mechanisms for the utilization of molecular oxygen. It is well known that the evolution of the physiology of utilization of molecular oxygen is closely related to the transition from procaryotic to eucaryotic cells. Structures of the latter, e.g., membranes, contain numerous substances whose synthesis requires O_2. For example, although sterols are not synthesized by bacteria, they are synthesized by various procaryotic blue-green algae, and this acquisition of function is believed to go hand in hand with the appearance of atmospheric O_2. We shall now begin to ask about the chemical origins of other structures, such as the nucleolus, that are uniquely present in eucaryotic cells.

Using enzymes of *E. coli,* the Tabors showed that two enzymes are essential in the biosynthesis of spermidine (425, 408, 419, 420). As presented in Fig. 15, the first enzyme decarboxylates S-adenosylmethionine (SAM), and a second transfers the aminopropyl moiety from decarboxylated SAM to putrescine to form spermidine and thiomethyladenosine. That the aminopropyl moiety of spermidine was derived from methionine had been demonstrated earlier (136, 51). Thus, we note that spermidine is derived from two important amino acids, ornithine and methionine, both of which are themselves under close metabolic controls. It is evident that methionine must be thought of as more than a moiety found in proteins or as a methyl donor.

The biosynthesis of spermine has been clarified only in the last two years. As shown by Pegg and Williams-Ashman (291, 293), animal cells contain an enzyme that appears to combine the two activities, i.e., decarboxylation of SAM and aminopropyl transfer (see Footnote 7). Furthermore, the rate of decarboxylation by the animal enzyme is positively controlled by the aminopropyl acceptor, putrescine. The same enzyme appears to be capable of transferring still another aminopropyl moiety to spermidine to form spermine. In animal cells both putrescine and spermidine stimulate the decarboxylation of SAM. Thus, some three centuries after Leeuwenhoek, the path of biosynthesis of the crystalline substance he observed in semen has begun to be clarified. We shall consider this matter in greater detail in the next chapter.

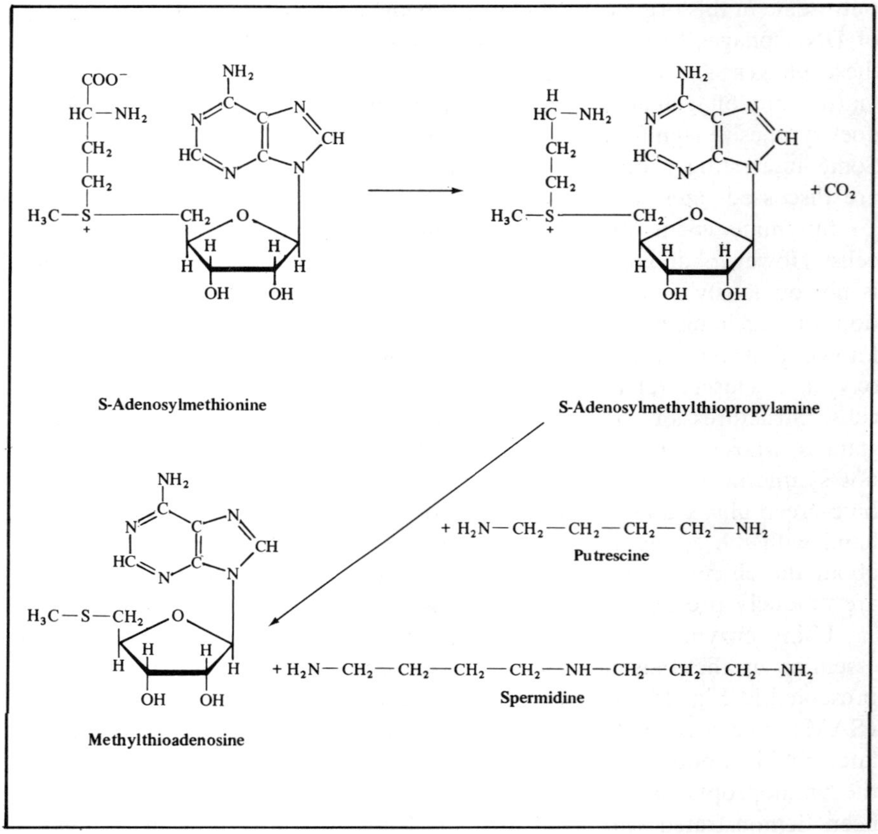

Fig. 15. The biosynthesis of spermidine from putrescine and S-adenosylmethionine.

The Functions of the Polyamines

What do these substances do? Are they essential to growth? These questions and the dissection of the molecular aspects of the role of these compounds are the subjects of the following chapters. However, it would be useful at this point to leave you with the awareness that the polyamines might be functionally significant. To that end, let us close by briefly considering data indicating that some polyamines are nutritionally important for a variety of organisms.

The first demonstration that a microorganism required these polyamines was made by Herbst and Snell in 1948 with *Hemophilus parainfluenzae*

(153, 152). The organism grew on a synthetic medium only when putrescine, spermidine, or spermine was added. Although 1,3-propanediamine was active, ornithine, ethylenediamine, and cadaverine were inactive as growth factors, as were many analogs of spermidine and spermine. The role of the growth factors, spermidine and spermine, has been obscured by the observed conversion of these substances to a diamine (446). Nevertheless, *Hemophilus parainfluenzae* has been used in initial approaches to the assay of polyamines in other cells and biological fluids (446).

These results were published several years after the work of Silverman and Evans showed that the inhibition of *E. coli* by atabrine or quinine was reversed by spermidine or spermine (370). It was suggested by these workers and by Snell (386) that these compounds and several other antibacterial and antiprotozoal substances interfere in an essential role effected by spermidine. More recently, it has been found that the substituted diamine, ethambutol, clinically important in the treatment of tuberculosis, inhibits the growth of mycobacteria in vitro and that this inhibition is reversed by addition of spermidine to the medium (110).

The growth of *Neisseria perflava* requires putrescine or spermidine (251). Figure 16(a) shows this requirement for putrescine. All strains of the fastidious organism, *Pasteurella tularensis,* appear to grow better in the presence of spermidine or spermine (436). These compounds facilitate the emergence of the organism from a lag phase, increase the rate of growth, and permit an increased density of growth. As in the case of *Pasteurella,* growth of lactobacilli from small inocula is also stimulated [Fig. 16(b)] by spermidine or spermine (141, 204). In addition, Mager showed that the lysis of *Neisseria* and *Pasteurella* and the leakage of material from the cells that occurred at low ionic strength were reduced by addition of spermidine (236). A leakage of RNA degradation products from *Hemophilus parainfluenzae* in polyamine-free medium was arrested by the addition of polyamine, and it was suggested that this was due to an inhibition of cellular nucleases (151). Thus, these results raise the question of whether the essential role of polyamines in these organisms is primarily that of stabilizing cell structures.

Several mutant stocks that are unable to synthesize putrescine have been obtained. The first of these was a mutant of *Aspergillus nidulans* obtained by Sneath (385) after ultraviolet irradiation. This organism grew maximally at 1 μg putrescine per ml. Spermidine replaces putrescine but is one-tenth as active at the same molar concentration; this was thought to be due to an impermeability of the mold to spermidine. Incubation in the absence of putrescine was reported to lead to chromosomal abnormalities. The hereditary difference between parent and mutant was determined

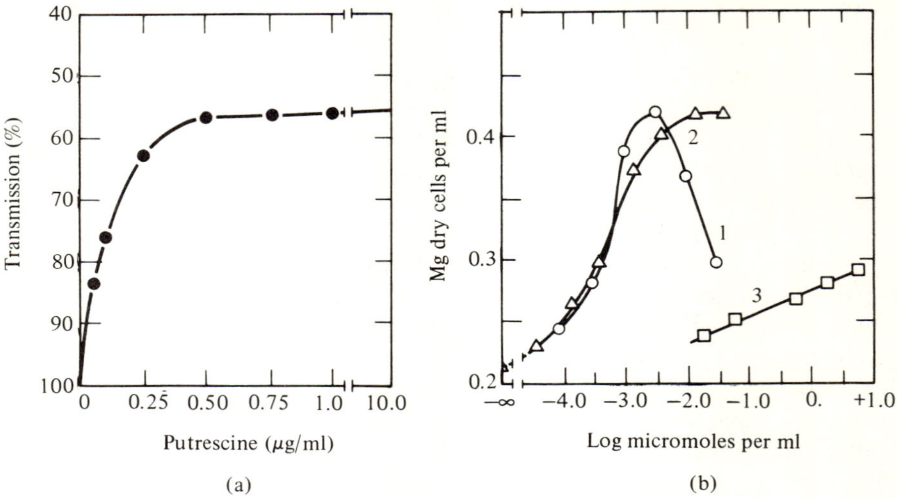

Fig. 16. Growth of microorganisms in the presence of exogenous polyamine. (a) Growth response of *Neisseria perflava* to different amounts of putrescine in a chemically defined medium (251). (b) Comparative effects of spermine, spermidine, and putrescine on growth of *Lactobacillus casei* (204). 1, Spermine; 2, spermidine; 3, putrescine.

by a single gene as shown by appropriate crosses.

Finally, we note the recent isolation by Maas and by Morris (169, 265) of putrescine-deficient mutants of *E. coli.* The organism isolated by Maas (169) was selected from canavanine-resistant cells. It was unable to convert arginine to putrescine, while the presence of arginine in the medium repressed and inhibited ornithine synthesis. A very slow rate of growth was observed in arginine-containing media in the absence of exogenous putrescine, and this was immediately converted to a more normal rate when putrescine or spermidine was added. Since *E. coli* does not convert spermidine to putrescine, it appears in this instance that spermidine may be the essential functional entity. The properties of these mutants are discussed in greater detail in Chapter 3. It is of interest that, with the exception of a single unusual *Pseudomonad* (see Footnote 3), no organism has yet been isolated that is incapable of synthesizing spermidine if putrescine is present in the medium.

Summary

We are concerned mainly with three rather simple organic cations, putrescine, spermidine, and spermine, whose structures have been known for some 45 years. The first two compounds are present in all procaryotic and eucaryotic cells. Spermine, although utilizable by procaryotes, is synthesized only by eucaryotes. It is just recently that the path of their biosynthesis has been clarified. However, only a few organisms unable to synthesize putrescine are known, and these grow slowly or not at all in its absence. In all cases, the requirement for putrescine can be replaced by spermidine, which restores growth. No organisms or mutants are known that are unable to synthesize spermidine from putrescine. The inhibitory effects of various chemotherapeutically useful drugs such as atabrine, quinine, and ethambutol are reversed by spermidine in selected systems. Enzymes exist that oxidize polyamines to highly toxic aldehydes, which may serve in their own right as antibacterial antibiotics.

Chapter 2

Polyamines in Eucaryotic Organisms

". . . Liquid Dew of Youth." It's already in production in a Leningrad vitamin plant. It's called Decamevit and has been referred to in the Russian press as "the elixir of youth and vigor." What is it? A combination of ten vitamins plus Methionin which is described as "a substance that furthers renovation of cells."

ATLAS, APRIL 1970

In the last chapter I called your attention to the cellular ubiquity of the organic cations, the polyamines, and stated that with very few exceptions all procaryotic cells contain putrescine and spermidine. These substances are frequently present in millimolar concentrations. Proof exists that the precursor of spermidine, putrescine, is essential for growth, but growth will ensue if spermidine is added to the putrescine-deficient organism. Since spermidine is frequently not transformed to putrescine, the inference may be drawn that spermidine may be the active agent essential to growth in several organisms.

Eucaryotic cells also contain a further derivative of spermidine, namely spermine, and, in the synthesis of spermine, a genuine acquisition of function unrelated to O_2 metabolism has appeared at this level of cellular evolution. The presence of this new compound has been accompanied in the evolution of some mammals by another enzyme, spermine oxidase, capable of converting spermine into a potential antibiotic, the dialdehyde, or oxidized spermine. There is suggestive, but not conclusive, evidence that spermine serves this function in certain extracellular fluids, such as seminal fluid. In any case, the effect of unchanged spermine (and spermidine) on eucaryotic cells is best studied in the absence of this enzyme.

In approaching the problem of cellular function, we shall briefly examine a few data concerning the amounts of the polyamines in various eucaryotic cells. Mention was made earlier of finding spermine in a eucaryotic alga (*Prymnesium*) grown in a synthetic medium. The compound has also been recorded as present in snake venom (435), sponges, sea urchins, insects, etc.

Polyamines in Animal Cells Grown in Culture

Cultured animal cells have been shown to contain putrescine at 0.25–2.5 mM, and spermidine and spermine at the level of 1–2 mM. Rapidly growing cells tend to contain more spermidine than spermine. The actual levels obtained will depend on the state of the medium. According to Pett and Ginsberg (297), fresh growth media cause a rapid synthesis of putrescine in KB cells; this synthesis requires protein synthesis, which is not supported in "spent" media. Ornithine present in the medium is converted to putrescine readily, but arginine is very slowly converted to this diamine. The putrescine in such cells can be converted to spermidine, which does not appear to revert to putrescine. Spermidine is converted to spermine, which can be degraded back to spermidine. These substances evoke sharp feedback inhibitions of putrescine biosynthesis from ornithine, as presented in Fig. 17.

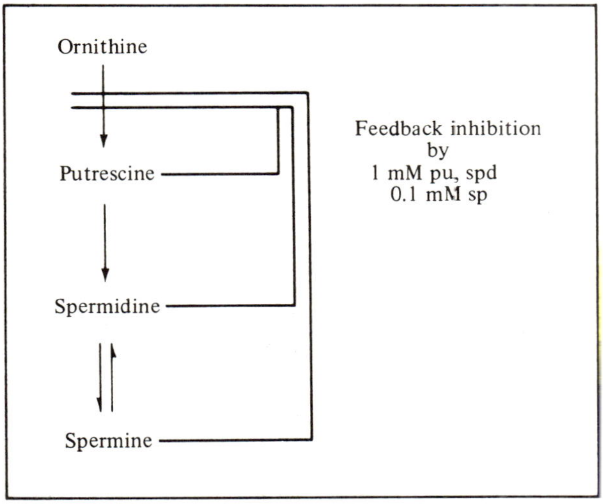

Fig. 17. The path of biosynthesis of polyamines in KB cells (297)

The early data compiled by the Tabors (421) on the concentration of spermidine and spermine in the tissues of various mammals suggests a rough correlation with the activity of the particular tissue in growth.

The relation of polyamines (in the developing rat) to growth potential is also evident in the studies of Raina and his collaborators. These workers showed that embryonic tissues are several-fold higher in spermidine than are older tissues (191). Aging of the tissue is accompanied by a marked decrease in the ratio of spermidine to spermine, due primarily to a sharp decrease in spermidine concentration (Fig. 18).

Shimizu et al. (362) have similarly observed marked decreases in the concentration of spermidine and spermine during the development of mouse brain. An increasing interest in the biogenic amines had led Shimizu and his colleagues to estimate these polyamines in the rabbit nervous system. Spermidine was markedly higher than spermine (10–20:1) in white matter and peripheral nerve, whereas spermine had increased in grey matter, decreasing the ratio to 2–5:1 (360). More recently, Kremzner has described extensive studies of polyamine distribution and metabolism in the nervous systems of many mammals (213).

The results of Shimizu thus suggested more specifically not only that spermidine may relate to active synthesis but also that spermine may be conserved in nuclei. He went on to compare the polyamines of mammalian

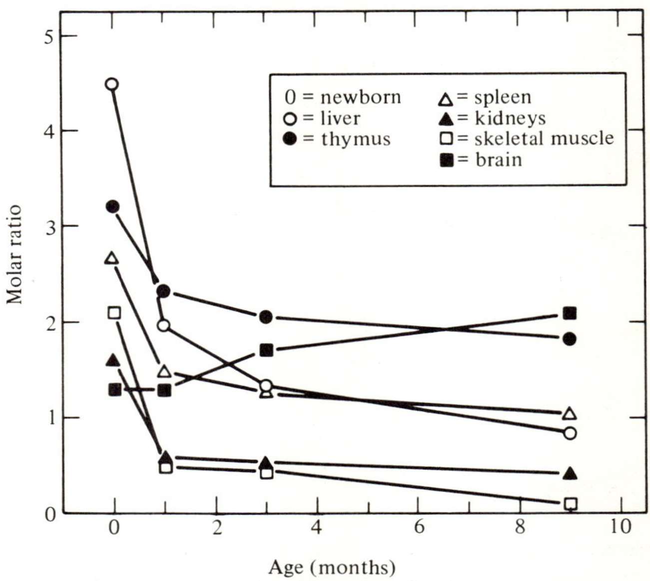

Fig. 18. Changes in the spermidine-to-spermine ratio in six rat tissues in relation to age (191).

erythrocytes and their nucleated counterparts in birds, etc. (361). Most mammalian erythrocytes are very low in polyamine; in fact, it is possible that much of the polyamine detected derived from nucleated leucocytes. You may recall that Charcot first observed crystalline spermine phosphate in blood rich in leucocytes. Nucleated erythrocytes generally have a much higher polyamine content, a datum pointing in the same direction, namely, that in these cells spermine may be largely a nuclear component. However, the surprising relatively high level of spermidine in rabbit erythrocytes cannot be explained in this way and indicates a new and unexpected problem concerning the role of spermidine in this cell.

Polyamines and Organelles

These crude correlations did not go entirely unnoticed and have elicited a few isolated studies on the distribution of the polyamines in animal cells and on the effects of exogenous polyamines on growth and other parameters in the systems.

Some 10 years ago, Lichtenstein and I (66) described the presence of a significant portion of cellular spermidine on the ribosomes of *E. coli;* we demonstrated that this substance was effective in associating 30 S and 50 S subunits to form the normal ribosomal monomer of 70 S. Since then, many workers have used spermidine in this way in replacing Mg^{++} in various studies of ribosome function. Our own conclusions were that "Mg^{++} and spermidine are the naturally occurring clasps on the complex ribosomal components existing in the bacterium."

These observations have been extended by many workers. However, as is discussed in a subsequent chapter, the actual significance of the presence of spermidine on a particular organelle or polymer may be obscured by the problem of the possible redistribution of polyamines during cell fragmentation and organelle isolation.

In any case, we can now look at some results, such as those of Raina and Teleranta in Fig. 19 on the distribution of polyamines and nucleic acids in the organelles of rat liver cells (311). The percentage distribution of spermine and spermidine in a rat liver homogenate resembles that of RNA in these experiments, in which the subcellular particles were isolated at relatively low salt concentrations (1mM $CaCl_2$). A slight excess of these polyamines over RNA is seen in the nuclear fraction; it may be asked if this excess implies some binding by DNA. The presence of significant amounts of both spermine and spermidine in nuclei of rat liver and calf thymus has been confirmed by Stevens (397), who has isolated the organelles by non-aqueous methods. Radioautographic procedures have been used by Dion and Herbst (86) to show that spermidine is preferentially

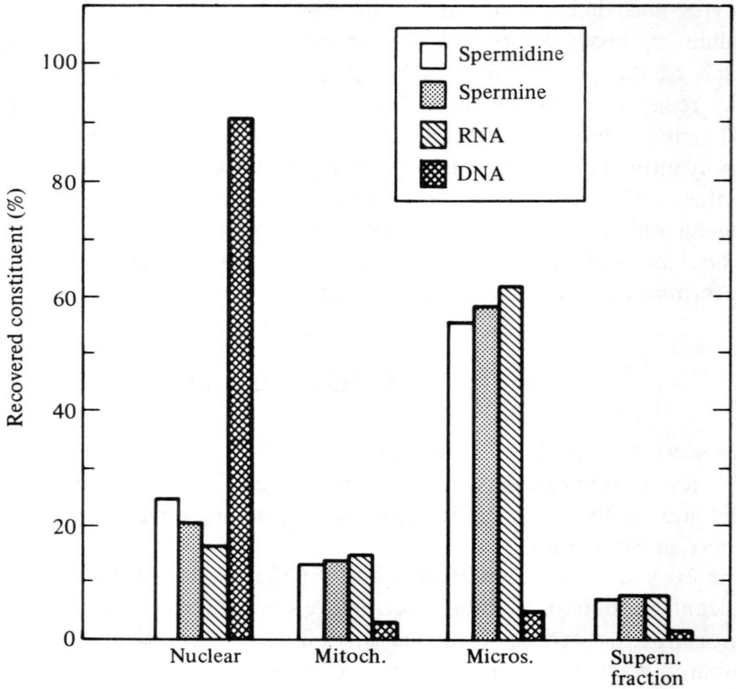

Fig. 19. Percentage distribution of polyamines and nucleic acids in subcellular particle fractions from rat liver (311).

associated with cell nuclei of isolated salivary glands of *Drosophila* larvae.

In Fig. 20 it can be seen that Ca^{++} or Mg^{++} causes the release of polyamines from a microsomal fraction, 50% of the spermidine being released at one-fifth of the concentration of divalent cation required for release of spermine, i.e., about 7 mM vs. 30 mM Mg^{++}. The binding therefore appears to be ionic, and the polyamines appear to have a much greater affinity for the particulates that does Mg^{++}, the cation generally thought to act as a stabilizer for ribosomal particles. The second part of this figure shows also that 50% of the spermidine is dissociated by about 60 mM K^{+}, whereas about 200 mM K^{+} is necessary to dissociate a comparable amount of spermine.

.In a recent study Khawaja and Raina (202a) have found that the attachment of ribosomes to the total membrane fraction of endoplasmic reticulum occurs only in the presence of 5 mM Mg^{++} or 0.3–0.5 mM spermine. Smooth membranes will not bind ribosomes whether spermine or Mg^{++} are present. Some laboratories have turned to the use of spermidine in the isolation of organelles. For example, in the isolation of metabolically active calf thymus nuclei in isotonic sucrose, Allfrey has found it advan-

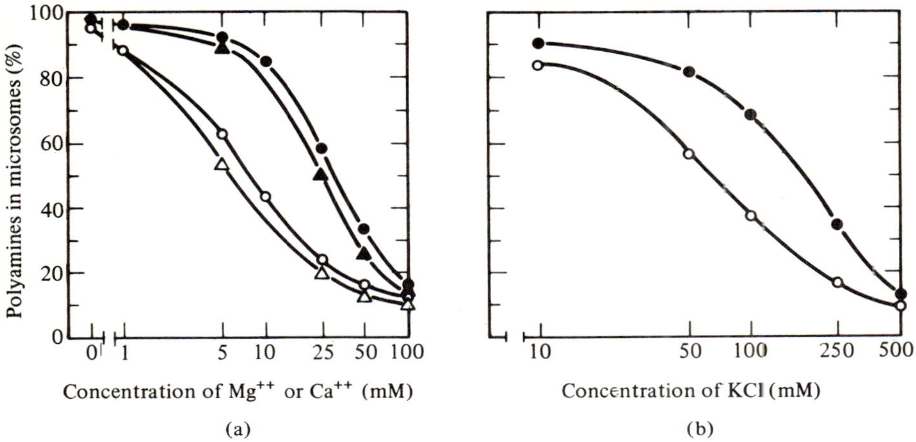

Fig. 20. Release of the polyamines from the microsomes into the supernatant by increasing cation concentration in the suspending medium (311). Total recovery (100%) was about 1 μmole spermidine, 1 μmole spermine, and 35 μmoles RNA-P. (a) Dissociating effects of Ca^{++} (triangles) (△ or ▲) or Mg^{++} (circles) (O or ●). Open symbols refer to spermidine, closed symbols to spermine. (b) Dissociating effects of K^{+}. Open circles (O) refer to spermidine, closed circles (●) to spermine.

tageous to replace 3 mM $CaCl_2$ by 1 mM spermidine (130). He has found that the nuclei produce large amounts of acetyl spermidine in such media (Allfrey, personal communication). Spermine also appears to stabilize liver nucleoli during isolation (44).

Brewer and Rusch have studied DNA synthesis in isolated nuclei derived from the mold *Physarum polycephalum* (41). As presented in Fig. 21, 0.6 mM spermine evokes a 10-fold stimulation of the incorporation of dATP. Of great interest is the fact that the addition of exogenous salmon sperm DNA to the spermine-containing system produces another 10-fold stimulation, i.e., almost 100 times the base level, whereas added DNA alone does not have this effect. Thus, the two together permit a better penetration of template to the nuclear DNA polymerase, or activate a latent polymerase, or possibly inhibit a nuclease. In fact, the peak activity of in vitro synthesis under such conditions is obtained at the time of the peak of DNA synthesis (and of thymidine kinase) in the mitotic cycle of the synchronously dividing organism (129). We have recently demonstrated millimolar levels of spermine, spermidine, and putrescine in *Physarum* at various stages of development.

The morphological effects of added polyamine on mitosis in plant root cells have been studied over the past decade. Treatments with 0.5 mM putrescine or spermine at different stages of division produce chromosomal abnormalities (9, 341, 80). In general, chromosomes are condensed within 3 hr and stay condensed, blocking mitosis irreversibly if treatment is main-

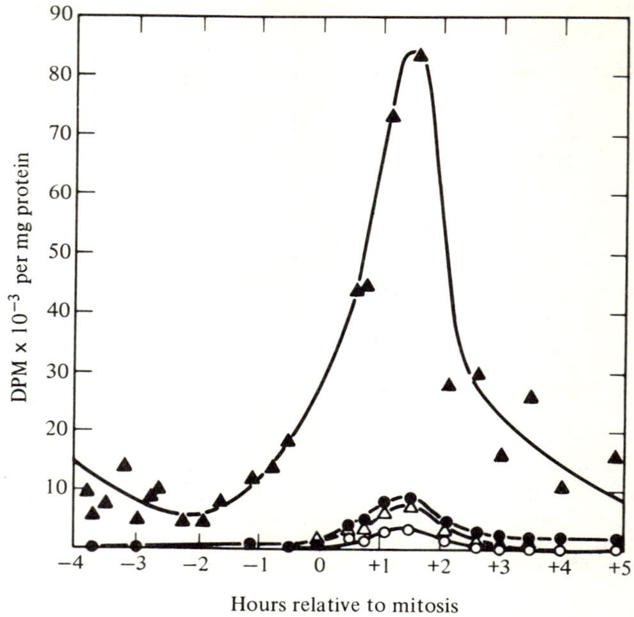

Fig. 21. Incorporation of ^{3}H-dATP by nuclei isolated at various times in the mitotic cycle (41). Nuclei were incubated in a standard medium with no additions (O), with added spermine, 5.7×10^{-4} M (●), with added DNA, 0.15 mg/ml (△), or with both spermine and DNA (▲).

tained for more than 6 hr. It has been suggested that naturally occurring polyamines may control chromosomal coiling and uncoiling.

On the other hand, 5 mM agmatine stimulates the mitotic rate of grasshopper neuroblasts in cultures of embryos. Also, the addition of relatively high concentrations of agmatine eliminate the mitotic delay observed after X irradiation of mouse fibroblasts (453). However, since agmatine is so difficult to purify and to maintain in that state, I wonder if this effect may not relate to some extent to the presence of other contaminating polyamines.

The condensing effects of 1 mM spermidine and spermine can be observed on isolated nuclei (80). The addition of these polyamines to relatively clear nuclei results in the appearance of chromosome-like condensates within the nuclei, but such structures have not been analyzed further.

Exogenous Effects of Polyamines

As noted earlier, a typical experiment in the exploration of function involves the addition of the polyamine. We must then attempt to eliminate the effects of toxicity produced by oxidized polyamine. In studies on the effect of spermine on a variety of mammalian cell lines, e.g., Ehrlich ascites cells and L-1210 leukemic cells in media supplemented by calf serum, toxicity was readily detected at very low concentrations, about 1 to 3 × 10^{-6} M, as noted in Chapter 1. In the presence of horse serum, which lacks spermine oxidase, toxicity disappeared, to reappear on addition of a small amount of calf serum or of a purified spermine oxidase. Nevertheless, the toxicity of spermine on KB cells in the presence of sera low in spermine oxidase suggests that toxicity in this system may not be related entirely to the presence of the dialdehyde or its degradation product, acrolein (161). Of course, we must consider the possibility that such toxic products may be generated intracellularly.

When Walker 256 cells were cultured in a medium supplemented with horse serum in the presence of 50 μg/ml spermine, i.e., a toxic level, the cells were clearly inhibited in the incorporation of tyrosine but stimulated in the incorporation of uridine (133). On the other hand, thymidine incorporation was unchanged under these conditions.

The growth of various animal cells has been studied in chemically defined media. In one such study, it was found that certain Chinese hamster cells could be grown in a synthetic nutrient mixture supplemented with two purified protein fractions, fetuin and serum albumin. Furthermore, such cells could multiply in such media in plastic petri dishes to form discrete colonies whose diameter was a function of cell growth and multiplication. It appeared that fetuin could be replaced by the polyamines under defined conditions, and, as shown in Fig. 22, the natural polyamines,

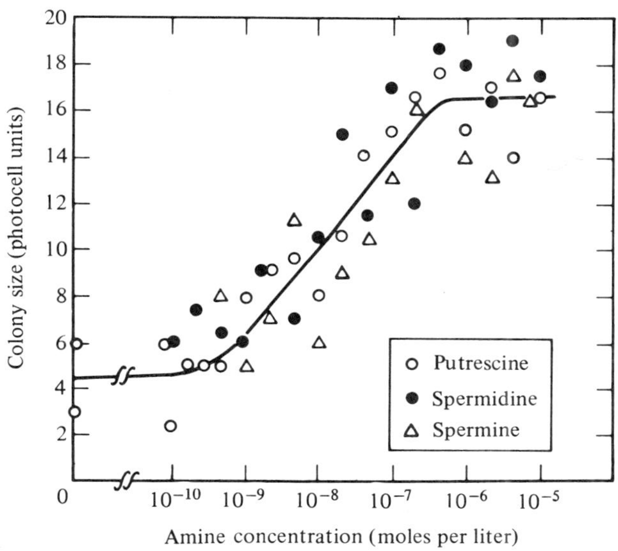

Fig. 22. Effect of putrescine, spermidine, and spermine on growth of strain CHD-3 cells. Incubation time = 12 days. Colony size was determined photometrically (145).

putrescine, spermidine, and spermine, were equivalently effective in facilitating cell extension, growth, and division for many generations (145).

Following our studies in *E. coli* on the stimulation of synthesis of ribosomal RNA by spermidine (69), Raina et al. have studied the effect of spermidine on incorporation of orotic acid or uridine in Ehrlich ascites cells incubated in buffer (310). At levels from 1–5 mM spermidine, 50% stimulations of RNA synthesis were regularly obtained. At concentrations of 10 mM spermidine and higher, incorporation was inhibited. It has also been observed that the presence of spermidine in an incubation of intact salivary glands of *Drosophila* larvae increases the incorporation of uridine into the RNA of cell nuclei (86).

It is relevant to our later discussions to note at this time that a number of chemotherapeutic agents that selectively block synthesis of ribosomal

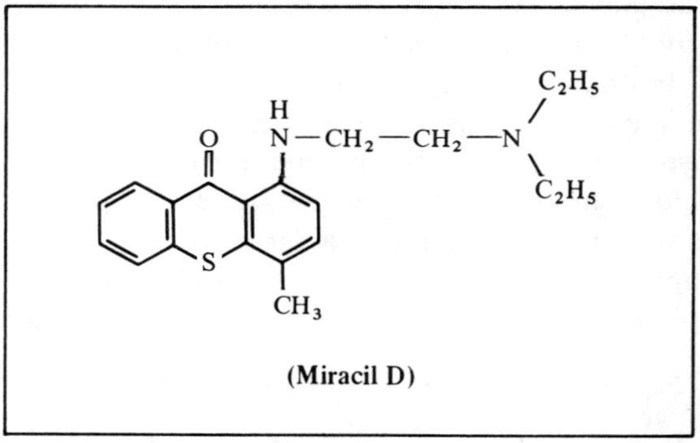

(a)

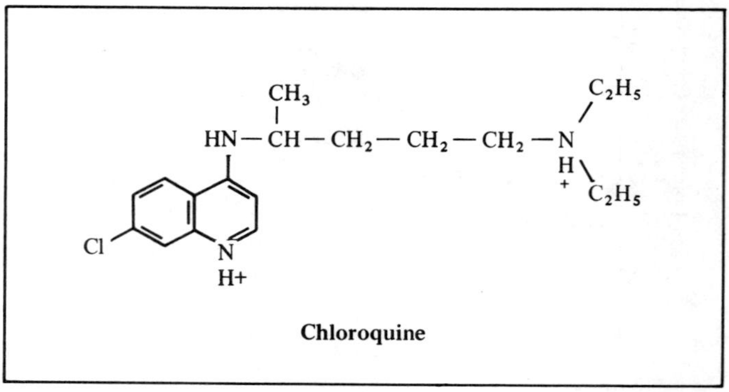

(b)

Fig. 23. Structures of therapeutically useful substituted diamines. (a) Lucanthone (Miracil D); (b) chloroquine.

RNA are substituted diamines (Fig. 23). Compounds like Miracil D, a potent schistosomicide (27), when added to HeLa cells in vitro, significantly reduced incorporation of uridine and thymidine into nucleic acids but did not reduce incorporation of amino acids into proteins (Fig. 24). Cell viability was reduced more rapidly than these biosynthetic functions. It has been shown by Bases and Mendez (27) that in short pulses of the drug for 2 hr, i.e., before loss of viability, there has been a potentially reversible selective inhibition of synthesis of ribosomal RNA, but not of synthesis of mRNA or of DNA. The structure of Miracil D, shown in Fig. 23, consists of an aromatic ring linked to a diamine, of which the latter is essential to activity (167). Binding studies of many such compounds, including chloroquine (Fig. 23), suggest that the diamine facilitates binding to nucleic acid and positions the aromatic ring system for interaction with the nucleic acid bases. We consider this hypothesis in greater detail in a subsequent chapter.

It was noted earlier that the presence of exogenous polyamine can protect some bacteria against autolytic degradation and prevents excretion of the resulting nucleic acid fragments (151). A similar result has been obtained with yeast strains suspended in media of low ionic strength; the RNA of the cells are then very sensitive to nucleases (348). Spermine at 10^{-6} M markedly inhibits the degradation by ribonuclease in non-growing cells. This concentration is not inhibitory in growing yeast, which in fact

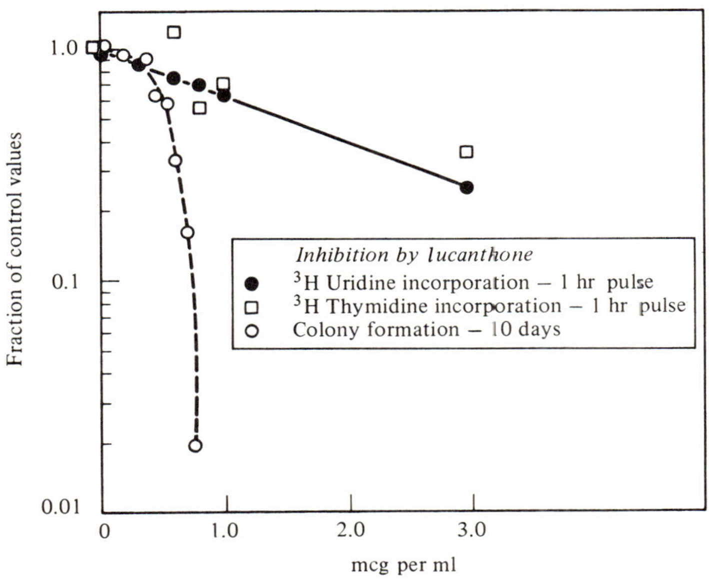

Fig. 24. The effects of Miracil D on RNA and DNA synthesis and on colony formation by HeLa-S-3 cells (27).

assimilates the polyamine. Under conditions of growth, much higher concentrations of spermine are required to establish inhibition (313, 343).

In the period after World War II, Rosenthal at the National Institutes of Health was among the first workers to begin studying the effects of the polyamines. His initial pharmacological observations demonstrated the relatively specific nephrotoxicity of spermine salts at levels of above 15 mg per kg in mice, rats, and rabbits (331). These results, in which renal tubules were seen to be badly damaged, led him and the Tabors into the more comprehensive analytical and metabolic studies that have provided the base on which current investigations of polyamine function have developed in many laboratories. Their early pharmacological work revealed that spermidine and spermine were synthesized in developing rats and maintained at carefully controlled levels in particular tissues, regardless of diet (332). However administered, spermine was concentrated in the kidney, whereas a small portion (4–21%) appeared in the urine at 24 hr with approximately equal amounts of spermidine, which arose as a degradation product. A portion of administered spermidine, which was far less toxic than spermine (331), was also excreted (332).

Several pharmacological studies have been undertaken in the past few years to explore the possible relationships among the polyamines, blood histamine levels, and psychological states. More careful studies of the toxicity of spermidine in mice have been made, and the effects of the compound have been observed on motor and nervous activity (120). De Meis has demonstrated that 0.5–2 mM concentrations of spermidine and spermine activate Ca^{++} uptake by muscle microsomes and inhibit microsomal ATPase (83). The compounds also inhibit actomyosin ATPase and relax glycerol-treated muscle fibers. It has been reported that spermidine is associated in significant measure with the fraction corresponding to nerve endings (259).

A recent comparison of histamine and spermidine contents in fetal rat brain during development has shown not only that the changes of both of these polyamines correlate with cell proliferation and growth during maturation, but also, as can be seen in Fig. 25, that spermidine follows the general pattern of histamine with an apparent lag of 24–48 hr (290). The precise significance of these correlations is not clear as yet, but it seems likely that in animals structural and functional interrelationships between the biogenic amines (histamine, serotonin, etc.) and the polyamines (spermidine and spermine) will be found. Of course, the aromatic moieties of the former will confer significant specificity in the effects of these compounds. It must be anticipated, however, that the latter, which are present in considerable excess, will competitively modulate the activity of the biogenic amines.

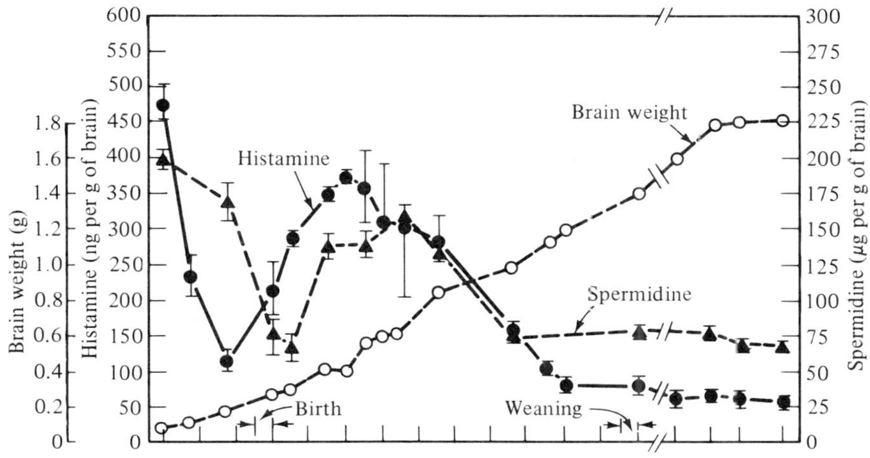

Fig. 25. Histamine and spermidine content in rat brain during development (290).

POLYAMINES DURING GROWTH AND DEVELOPMENT

In 1960 Raina began his important studies at the University of Helsinki on the polyamines of vertebrate tissues. After developing a new useful assay for spermidine and spermine that depended on amido black staining of electrophoretically separated polyamines, he examined the developing chick embryo (303). With isotopically labeled precursors, he has demonstrated the utilization of ornithine, putrescine, and methionine to form spermidine and spermine (302). It can be seen in Fig. 26(a) that in the utilization of ^{14}C-putrescine, the formation of spermidine considerably precedes incorporation into spermine. In the utilization of 2-^{14}C-methionine, however, the two curves are more similar [Fig. 26(b)]. These results are consistent with the addition of the aminopropyl moiety of methionine to some preexisting unlabeled spermidine. Raina also showed a utilization of ornithine comparable to that of putrescine and concluded that a pathway of spermidine synthesis similar to that existing in bacteria was present in this vertebrate. This work on the biosynthetic pathway in the chick is a natural development of earlier studies on biosynthesis in procaryotic and eucaryotic microorganisms.

During chick development there is a very extensive, greater than 50-fold, net synthesis of the polyamines until just before hatching [Fig. 27(a)]. Nevertheless, the relatively enormous initial concentration of spermidine (10 μmoles per g dry weight, 5 μmoles per g wet weight) falls sharply (about 50%), attains a relatively high plateau level, and then falls to the

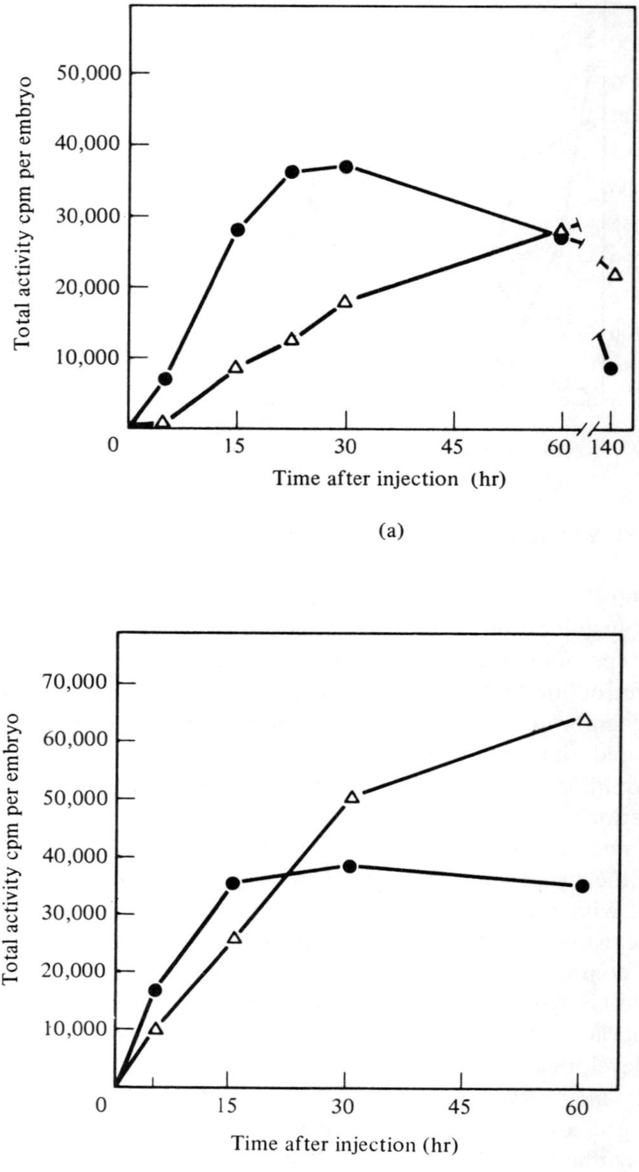

Fig. 26. Incorporation of radioactivity into spermine (△) and spermidine (●) in the developing chick embryo after injection of (a) 1,4-^{14}C-putrescine and (b) 2-^{14}C-methionine onto the chorioallantoic membrane (302).

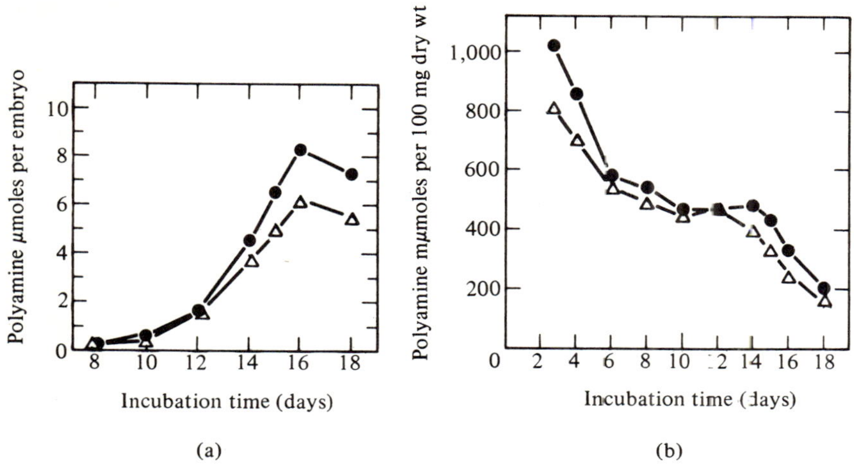

Fig. 27. (a) Total spermidine and spermine content (μmoles/embryo) of the chick embryo in relation to age (303): ● = spermidine; △ = spermine. (b) Polyamine content, expressed as mμmoles per 100 mg dry weight.

adult level (1–2 μmoles per g) at 18 days [Fig. 27(b)]. At this time, some tissues contain more spermidine than others. Liver is very high, whereas heart muscle is relatively low, i.e., less than one-third that of liver. The molar ratios of spermidine to spermine also varies in these tissues, liver being 1.17 and heart being 0.47. Thus, the same general correlation exists, implicating these polyamines in growth and protein synthesis.

In Bologna a group led by Caldarera has more recently studied the developing chick and extended these results. This group has demonstrated an active polyamine-degrading activity, ascribed to spermine oxidase, induced by injection of spermine into the yolk sac (45). After this induction, spermine and spermidine fell, with a concomitant remarkable decrease in RNA and DNA. When this degradation was inhibited by injection of iproniazid, a known inhibitor of the amine oxidases, the polyamines rose to abnormal levels with a parallel increase in the nucleic acids (45). The polyamines are thus shown to have a more active metabolism in the chick than had been anticipated. Moreover, they correlate even more closely with the nucleic acids. The fact that their apparently specific increase or decrease, effected enzymatically, produces parallel changes in the nucleic acids suggests that the concentration of polyamines may actually control that of the nucleic acids, directly or indirectly.

Probing more deeply, this group has shown that an increase in polyamine concentration caused by injection of iproniazid increases incorporation of orotic acid into RNA and DNA, particularly into ribosomal RNA and nuclear DNA (271). The activities of DNA-dependent RNA polymerase were similarly enhanced. Under these conditions larger polysomal aggregates were also formed, a result suggesting that the polyamines may have a role in the control of protein synthesis.

Both the Japanese and the Italian groups have studied the polyamines in developing brain. As noted earlier, Shimizu observed that the concentration of polyamines in mouse brain was higher in prenatal and early postnatal life than in the adult. Caldarera, continuing work with the brain of the developing chick, has obtained a similar result. In addition, he has found a natural increase of spermine oxidase during development, as shown in Fig. 28, and he suggests that the increase of spermine oxidase may in fact account to a significant extent for the limitation of brain development (48). Injection of spermidine or spermine into the air space of the embryonated egg on the 10th day led to a great increase in a few hours in formate incorporation into the RNA of ribosomes with 100–150% increments in total ribosomal RNA. Furthermore, as shown in Fig. 29, it was possible to find a great increase in polyribosomal aggregates (48).

Barros and Giudice, working with sea urchin embryos, have also reported that the synthesis of ribosomal RNA from the stage of fertilization to early gastrulation can be stimulated by addition of spermidine to sea water (26).

Caldarera et al. have recently described the changes in polyamine during embryonic and larval development of the frog (46). Spermine, invariably present in the adult liver, is very much lower in the embryonic and larval animal. The concentrations of putrescine and spermidine increase sevenfold from the unfertilized egg to the beginning of larval development. Larval development begins with an initial drop in polyamine, which may cor-

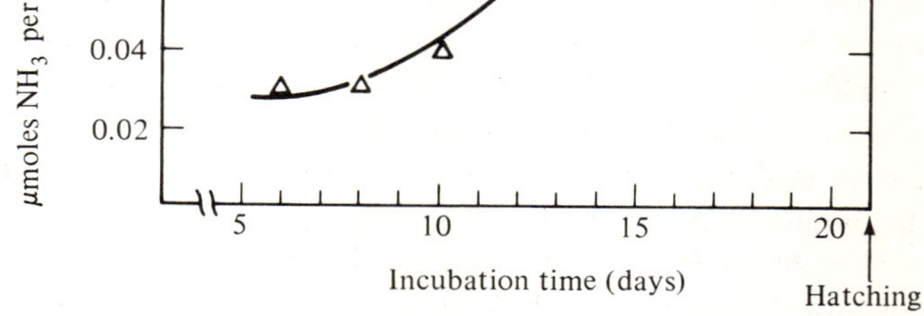

Fig. 28. Spermine oxidase activity during development of the brain of the chick embryo (48).

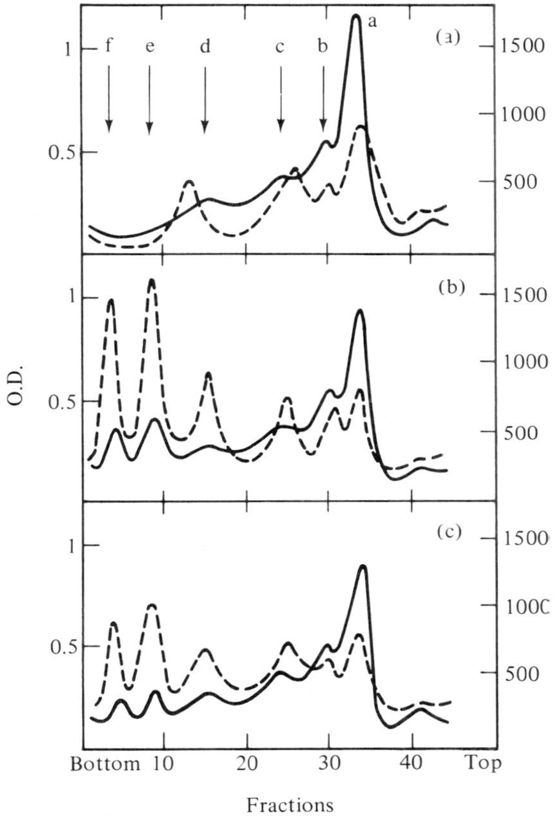

Fig. 29. Effect of spermine or spermidine injection on incorporation of ^{3}H-formate into polyribosomal aggregates (48). The embryos were injected with polyamine (0.5 μmole in 0.3 ml) into the air space on the 10th day of incubation. Two hr later they were similarly injected with sodium ^{3}H-formate and harvested and homogenized after an additional 2 hr. Graph (a), injected with NaCl; graph (b), injected with spermine; graph (c), injected with spermidine (peak a = 73 S; b = 113 S; c = 147 S; d = 206 S; e = 259 S; f = 280 S). Solid curve, O.D. 260; dashed curve, radioactivity.

relate with the appearance of spermine oxidase. The concentrations of the polyamines are subsequently restored.

Using the sensitive dansyl method, Herbst and Dion (150) have described in some detail the pattern of polyamine changes during the development of *Drosophila melanogaster*. Fertilization of the insect egg evidently produces a rapid conversion of the putrescine pool to spermidine. The very high concentration of this substance at 0.5 hr then drops in a few hours to a level that is maintained until hatching at 22–24 hr, and decreases only slightly at subsequent larval stages. Putrescine remains low in concentration but is in fact being actively synthesized and incorporated into spermidine. The total amount of spermidine increases markedly at different stages simultaneously with RNA synthesis and growth. These authors also consider that spermidine participates actively in the regulated synthesis of nucleic acid and protein.

Polyamines in Hypertrophy

A single intravenous injection of folic acid in the rat stimulates renal growth with a concomitantly large increase in RNA and DNA synthesis. Kremzner and Cote (214) have shown that the content of spermidine in the rat kidney decreased slightly in the first 3–5 hr, followed by a sharp increase to double the spermidine content in 48 hr. This closely parallels the net synthesis in RNA. Experiments in this system with ^{14}C-putrescine indicate an increase in putrescine uptake and increased incorporation into spermidine.

When nephrosis is induced in the rat by injection of antikidney serum, with loss of serum protein, there is an increased hepatic synthesis of plasma protein, accompanied by an increase in RNA, DNA, and liver weight. Marsh and his collaborators (248) have used a moderate nephrosis to effect a relatively small synthesis of DNA and RNA. It can be seen in Fig. 30 that such an increase is preceded by a sharp increase in liver spermidine.

Polyamines and the Synthesis of Macromolecules in Regenerating Liver

The relation of polyamines to the synthesis of DNA, RNA, and protein has been explored quite extensively in the study of liver regeneration, a classical system of rapid synchronized tissue development. In the work of Dykstra and Herbst in 1965 (95), the concentration of spermidine, but not of spermine, was observed in rat liver to increase within 24 hr after partial hepatectomy, and the rate of polyamine accumulation approximated the increased rate of synthesis of RNA in regenerating liver. The uptake of intravenously injected putrescine and the biosynthesis of spermidine was accelerated within 2 hr after the operation.

At this time Raina and his collaborators also turned to the study of rat liver. They showed that the incorporation of isotopic methionine into spermidine (Fig. 31) is stimulated soon after partial hepatectomy (308). On the other hand, the incorporation of isotope into spermine occurred rather late. These studies indicated that the ratio of polyamine-N to RNA-P

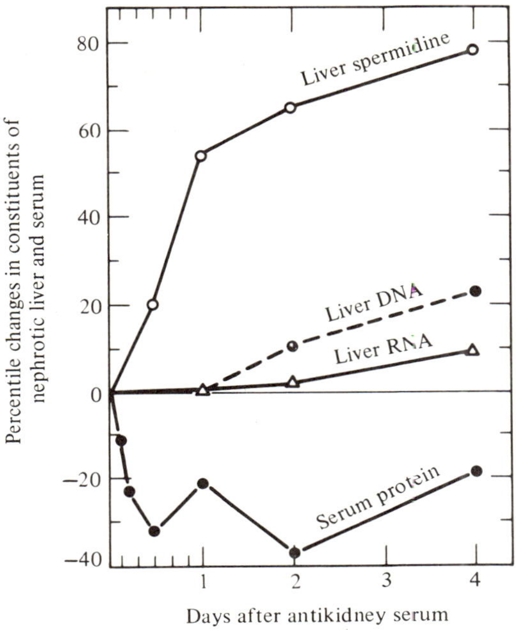

Fig. 30. Percentile changes in levels of spermidine, DNA, and RNA of liver and in total protein of serum, following administration of antikidney serum (248). The control point (0 hr) is the mean of 12–24 experiments; the other points are the means of 3–9 experiments.

remained approximately constant (308), a result that led Raina to suggest that the polyamines may serve to stabilize RNA.[4]

By now the close relation of polyamine-N to RNA-P has been observed many times in many systems, including bacterial systems discussed in Chapter 3. The rigorous stoichiometry of these functions is very well exemplified in the more recent studies of Seiler et al. (352) on mouse livers. The livers were treated so that RNA (Fig. 32) was either decreased by starvation or increased by drugs such as phenobarbital or perfluorobutyric acid. A change of RNA-P by 15 or 16 μmoles was accompanied by a change in spermidine by 1 μmole. Thus, spermidine-N accounts for 20% of the RNA-P.

In extending the work of the Finnish group, Jänne studied the biosynthesis of the polyamines in liver of developing rats in some detail

[4] In a recent paper, it has been shown that the half-life of spermidine in normal and regenerating rat liver is four days (335a), approximating a half-life obtained for ribosomal RNA in normal liver. The half-life of putrescine in mammalian organs is of the order of two hr (335a).

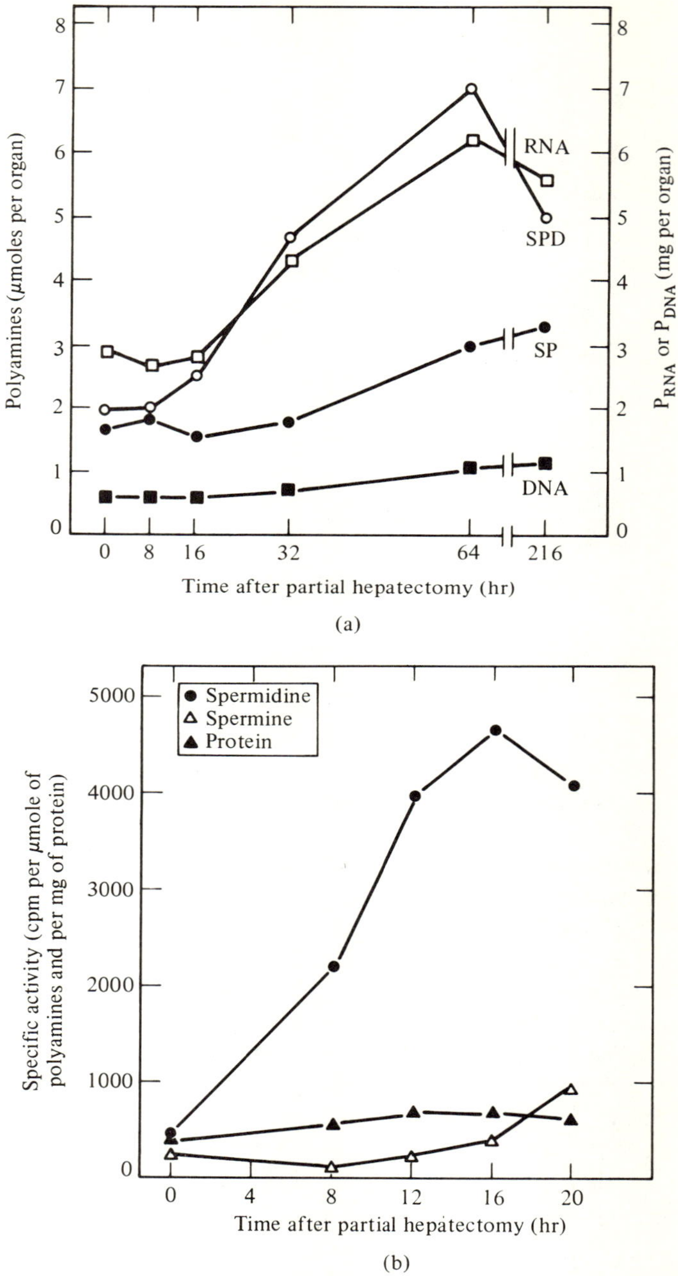

Fig. 31. (a) Total polyamine and nucleic acid contents in rat liver after partial hepatectomy (308). SPD = spermidine; SP = spermine. (b) Incorporation of radioactivity into liver polyamines and total protein in partially hepatectomized rats after intraperitoneal administration of 2-^{14}C-methionine (308). Injection time was 4 hr before analysis.

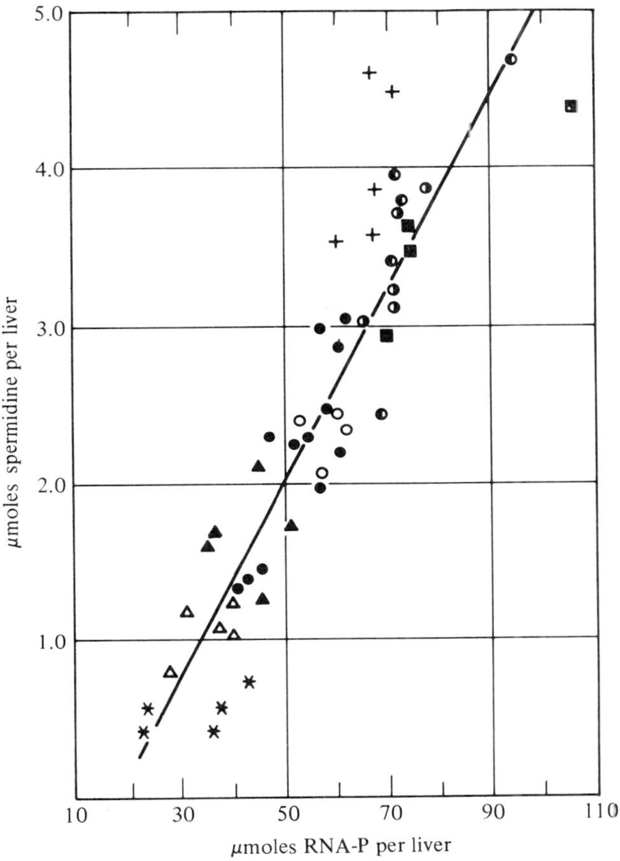

Fig. 32. The variation of spermidine and RNA in mouse liver under various conditions (352). Control (●); 36 hr starvation (△); 36 hr starvation + phenobarbital (▲); 48 hr starvation (✱); 48 hr phenobarbital (O); 88 hr phenobarbital (◐); 112 hr phenobarbital (◑); 168 hr phenobarbital (■); perfluorobutyric acid (+).

(187). He demonstrated particularly the formation and utilization of putrescine and the rapid degradation of much injected diamine to CO_2. The conversion of putrescine to spermidine was inhibited by administration of ethionine to the rats.

The surprising result was also obtained that the specific activity of spermidine after injection of putrescine was no greater in sham-operated liver than in regenerating liver (187, 188). It was shown that the concentration of liver putrescine doubled within 4 hr after partial hepatectomy, a result of an increased rate of biosynthesis from ornithine and arginine. It appeared that the increase in biosynthesis of spermidine in regeneration is due to an initial increase in putrescine biosynthesis and accumulation (187).

Kostyo showed that hypophysectomy of the rat led to a marked reduction of spermidine concentration (about 50%) in the liver. Injection of growth hormone into such animals produced a return of spermidine, RNA, and protein synthesis in 24 hr, without any change in the spermine level (208).

Jänne tested the effects of growth hormone on normal liver. The time course of synthesis of net spermidine from methionine was similar to that for RNA and protein. However, the specific activity of the spermidine had increased sharply, as seen in Fig. 33. An early increase was observed in the conversion of ornithine to putrescine; a several-fold increase in ornithine decarboxylase was detected (187, 192). The decarboxylase was not increased in animals treated with actinomycin or puromycin. Putrescine actually accumulated; however the capacity of normal liver to convert exogenous putrescine to spermidine did not increase. Although it was thought that the stimulation in spermidine synthesis after hormone treatment stemmed primarily from increase in the concentration of putrescine, it is not yet evident whether there is a marked increase in the enzymes of spermidine synthesis in these systems.

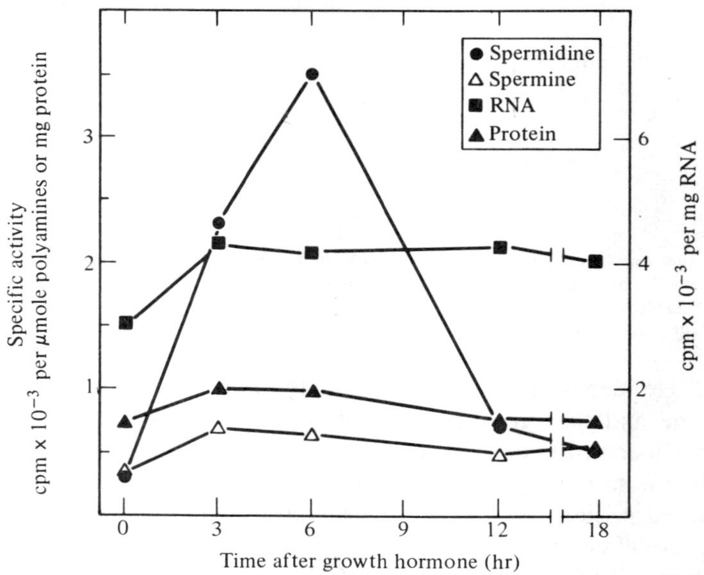

Fig. 33. Effect of growth hormone on the synthesis of polyamines, RNA, and protein in rat liver (192). Two-month-old rats received growth hormone and 2-^{14}C-methionine or 6-^{14}C-orotic acid 30 min before sacrifice. Controls were treated with the solvent alone. The radioactivity of the protein refers to treatment with ^{14}C-methionine.

The observation that ornithine decarboxylase is increased during growth had been demonstrated initially by Jänne in regenerating liver (142). In this system, however, a 40-fold stimulation was observed in contrast to the hormone-stimulated "normal" liver (189). The enzyme was purified (306) and shown to be different from other decarboxylases such as decarboxylases for L-glutamic acid and L-tyrosine, which were not significantly increased after partial hepatectomy. The ornithine decarboxylase of regenerating rat liver has a slightly alkaline pH optimum, an absolute requirement for pyridoxal phosphate, and is not significantly inhibited by 10 mM putrescine, agmatine, or spermidine (306).

This work has been extended by many workers in the United States. In particular, Russell and Snyder have been the most active. They have confirmed that ornithine decarboxylase is specifically stimulated in regenerating rat liver (336). This stimulation does not occur in a hypophysectomized animal, except in the presence of growth hormone (338). Even as described by Jänne, single injections of growth hormone produce relatively enormous increases of the enzyme in adult and weanlings. The reaction in the latter is even greater than in the adult.

Of great interest is the finding that in fetal rat liver and some tumors there is relatively high histidine decarboxylase in contrast to the high ornithine decarboxylase found in chick embryo and some other tumors (336). One wonders if, in the fetal system, histamine does not fulfill a role comparable to that of spermidine in regenerating liver.

We should mention that the purified polypeptide from the submaxillary gland that stimulates growth of epidermal tissues also produces in such tissues a specific 40-fold increase in ornithine decarboxylase, followed by accumulations of newly synthesized putrescine and spermidine (392). Schrock et al. (349) have reported that many treatments stimulating RNA synthesis and growth in rat liver increase this decarboxylase specifically (349). Some stimuli, however, not affecting growth also increase the enzyme, which does not appear linked to DNA synthesis.

Fausto has shown that thioacetamide, a compound that stimulates RNA synthesis in normal liver, also stimulates synthesis of this enzyme in normal animals (105). In addition, he has reported that injection of thioacetamide in normal rats causes an increase of liver spermidine and an increase in the synthesis of nuclear RNA (106). It is possibly very significant that a major cytological effect of thioacetamide (393) is that of enlarging liver nucleoli, the site of synthesis of ribosomal RNA.

It has been shown (190) that stimulating the formation of ornithine decarboxylase in the presence of growth hormone causes a less striking but nevertheless significant increase in RNA polymerase (Fig. 34). This increase of 100–150% does not occur in the presence of cycloheximide plus growth hormone so a de novo synthesis of the polymerases is suggested.

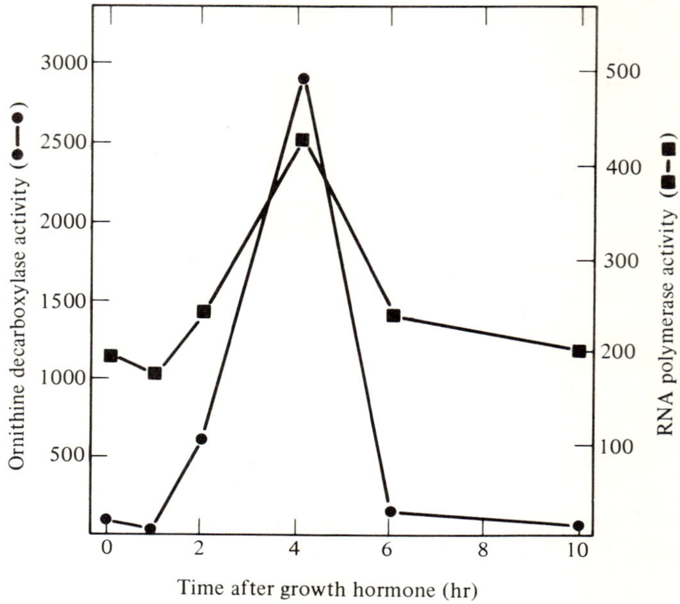

Fig. 34. Time course of the effect of growth-hormone treatment on levels of RNA polymerase and ornithine decarboxylase in rat liver (190).

However, since the RNA polymerases in nuclei and nucleoli are stimulated by spermidine and spermine, as is seen later, we wonder if this relatively small increase in the polymerase does not arise to some extent from the increased amounts of the polyamines in the homogenates.

Russell and Snyder have demonstrated an extraordinary turnover rate for ornithine decarboxylase (337). Figure 35 describes how the administration of cycloheximide or puromycin to normal rats or to those with regenerating liver produces a rapid decline in the activity of the enzyme with a half-life of about 11 min. This is not due to an instability of the isolated enzyme or an inactivation in homogenates. Not only is this turnover extremely fast, but the result strongly suggests that the increase in the enzyme after partial hepatectomy is due to synthesis of new enzyme and not to a decrease in the rate of inactivation. Thus, as can be seen in Table 1, this enzyme appears to have the most rapid turnover rate of any mammalian enzyme studied to date (337). It is difficult to believe that this phenomenon is not tied to the coarse and fine controls of growth initiation in liver.

Actinomycin D, which inhibits the synthesis of RNA by DNA-dependent RNA polymerase, completely blocks increase of the ornithine decarboxylase when the inhibitor is administered at hepatectomy (337). The block in enzyme increase is only partial when the compound is injected at 30 min

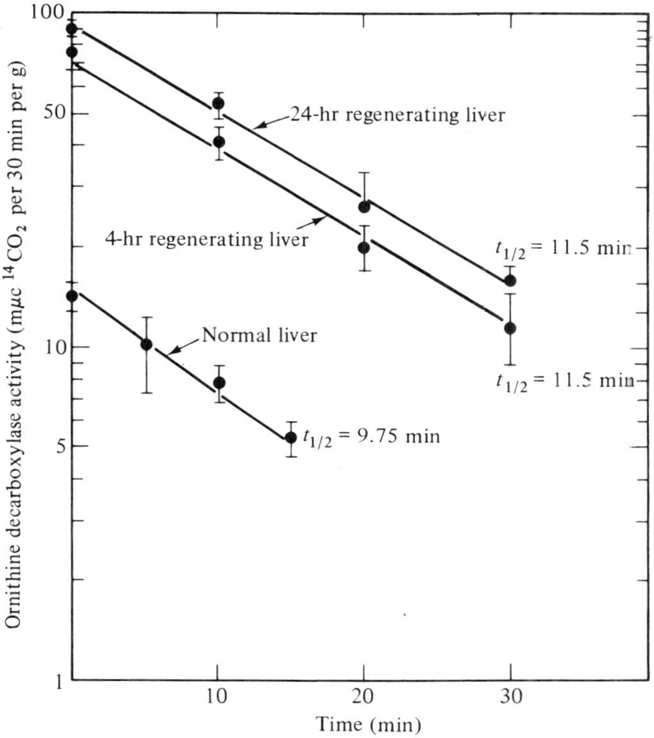

Fig. 35. Time course of the decline of ornithine decarboxylase in 4 hr regenerating rat liver, 24 hr regenerating rat liver, and normal rat liver after administration of 50 mg/kg of cycloheximide intraperitoneally (337).

Table 1. Half-Lives of Some Mammalian Enzymes (337)

Enzyme	$t_{1/2}$
Ornithine decarboxylase	11 min
δ-Aminolevulinic acid synthetase	67–72 min
Tyrosine transaminase	1.5 hr
Histidine decarboxylase	1.8 hr
Ribonucleotide reductase (conversion of CDP to dCDP)	2.0 hr
Tryptophan pyrrolase	2.5 hr
Threonine dehydratase	3.0 hr
Serine dehydratase	5.2 hr
Catalase	1–1.5 days
Glutamic-pyruvic transaminase	3.5 days
Arginase	4–5 days

and is non-existent if actinomycin D is injected at 1 hr. New species of RNA have been detected in mouse liver during the first hour after hepatectomy (57).

Some Paths of Ornithine Biosynthesis and Transformation

In Fig. 36 we can see a comparison of some differences in the origin of ornithine from glutamate in the mammal and in *E. coli.* Whereas both organisms make arginine via citrulline from ornithine, in mammalian liver the urea cycle generates ornithine from arginine. Thus, ornithine is the main precursor to putrescine from endogenous ornithine or exogenous arginine. However, in *E. coli,* which lacks urease, exogenous arginine is mainly decarboxylated to agmatine, which is then converted to putrescine by means of agmatine ureohydrolase. This latter route has not yet been found to be a major route in the mammal, although paths like it are found in plants. This route is considered in greater detail in Chapter 3.

Although the level of activity of aspartate transcarbamylase is closely regulated by the exigencies of nucleic acid synthesis, ornithine carbamoyl-

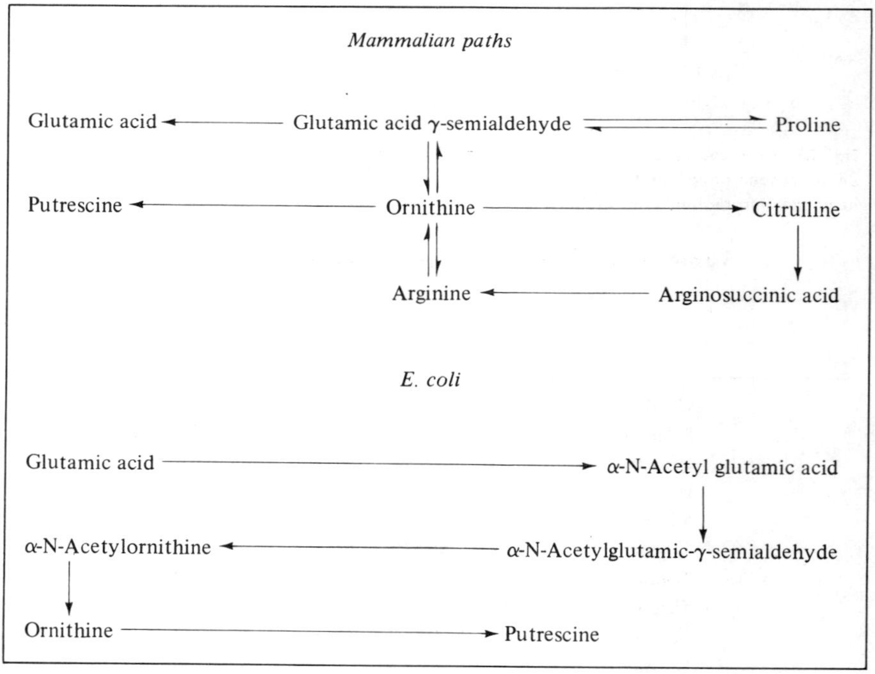

Fig. 36. Some paths of ornithine biosynthesis and transformation in the mammal and in *E. coli.*

transferase appears to be less variable than the decarboxylase. Thus, unlike the aspartate transcarbamylase and ornithine decarboxylase, the level of the ornithine carbamoyltransferase is similar in normal and regenerating liver (131).

On the other hand, the liver γ-ornithine ketoacid aminotransferase, which converts ornithine to glutamic acid-γ-semialdehyde, is relatively low in the fetal animal and is depressed following partial hepatectomy or repeated injection of growth hormone (301). This enzyme thus shows an inverse relation to the ornithine decarboxylase, and the balance of these two enzymes appears at first sight to be set to facilitate putrescine production after hepatectomy. Although the level of the transferase has been explored in a number of nutritional and genetic disorders in man (201) and as a function of hormonal factors (158), the state of ornithine decarboxylase and the level of putrescine accumulation in these situations has not yet received simultaneous study.

On the Metabolism of Spermine

Siimes has made a detailed study of the biosynthesis and degradation of spermine in rat tissue (364). Labeled spermidine was shown to be a direct precursor of spermine, as inferred earlier from studies with other precursors. Spermine is formed most actively in tissues in which the ratio of spermidine to spermine is the lowest. On the other hand, exogenous labeled spermine is also converted to spermidine; this conversion occurs most actively in tissues in which the ratio is highest. Thus, this ratio relates both to the biosynthesis and the degradation of spermine. As noted earlier, the appearance of isotopic carbon in rat liver into spermine from putrescine was significantly slower than that from methionine. In regenerating liver, as presented in Fig. 37, the formation of spermidine from spermine is seen to be more rapid than the converse. The liver is also capable of degrading spermidine to free putrescine.

The rates at which the polyamines are incorporated and degraded depended on the routes by which they are administered. The observed interconversions appear to be unaffected by the intestinal microorganisms. Administration of ethionine, which ties up ATP in S-adenosylethionine and reduces levels of S-adenosylmethionine, extensively inhibited the conversion of exogenous spermidine to spermine without significantly affecting the conversion of spermine to spermidine. During starvation, the concentrations of both polyamines was maintained for several days and then fell, although the amounts in the total liver tended to fall from the beginning. The rate of conversion of ^{14}C-spermine to spermidine is increased somewhat during starvation.

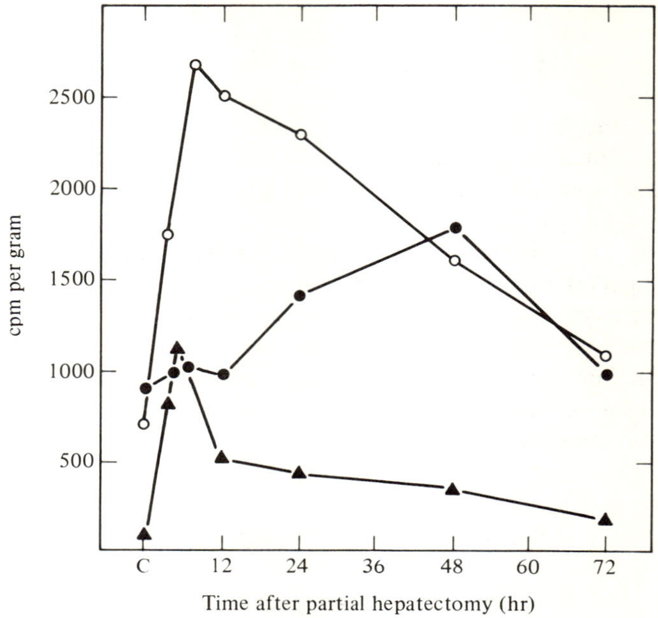

Fig. 37. Formation of spermine (●) and putrescine (▲) from 1,4-^{14}C-spermidine, and spermidine (O) from 1,4-^{14}C-spermine in regenerating liver (364).

Polyamine Biosynthesis in the Prostate

It is appropriate that the clarification of the route of biosynthesis of spermidine and spermine in the mammal should be attained in significant measure by study of the prostate gland. This gland is the major source of the spermine found in seminal fluid by Leeuwenhoek. The definitive work on this mechanism has been done by Pegg and Williams-Ashman (291) as recently as 1968.

Williams-Ashman has made many important contributions to the biochemistry of the prostate gland. He and Rhodes (317) demonstrated that in the rat the various regions of the gland are functionally distinct, the ventral and lateral lobes of the prostate (Table 2) being the major source of the large amount of spermine that appears in seminal fluid. Although the fluid is relatively low in spermidine, the ventral prostate contains 5–6 μmoles per g of both spermidine and spermine, and only one-tenth as much putrescine. The coagulating gland can be seen to be relatively low in polyamine.

As is well known, orchiectomized animals respond actively to injections of testosterone. In 1964 Williams-Ashman and his coworkers showed that the blood levels of androgenic sterols control nucleic acid metabolism in

Table 2. Polyamines in Male Accessory Reproductive Organs of the Rat (317)

Organ	μg base per 1 g wet weight of tissue	
	Spermidine	*Spermine*
Seminal vesicle	64, 159	98, 95
Seminal vesicle secretion	0	0
Dorsolateral prostate	410	570
Ventral prostate	720, 785	1120, 1170, 1106
Coagulating gland (anterior prostate)	63, 78	34, 67

Tissues pooled from Sprague-Dawley rats weighing 300–350 g

the prostate gland (456). With growing realization of the relation of the polyamines to RNA metabolism, this group went on to study the enzymatic mechanisms of polyamine synthesis in this gland and the controls on these mechanisms.

Concomitantly, Caldarera's group demonstrated not only that orchiectomy leads to a great drop in polyamine content in the prostate but also that testosterone maintains the original levels of these substances (47). Furthermore, they showed that the addition of polyamine to the isolated prostate nuclei of orchiectomized animals markedly stimulated both the Mg^{++}-activated and the Mn^{++}-activated RNA polymerase (Fig. 38). After orchiectomy, the fall of the Mg^{++}-activated enzyme [Fig. 38(a)] was greater than that of the Mn^{++}-activated enzyme [Fig. 38(b)].

It has been reported that the RNA product of the Mg^{++}-activated polymerase in isolated nuclei has a composition resembling a ribosomal RNA; i.e., this RNA is a product of nucleolar activity. At least the isolated nucleoli of rat liver do in fact produce a high guanine-cytosine type RNA in a reaction stimulated by Mg^{++} (174) and by spermidine (304, 307).

The drop in activity of the Mn^{++}-activated enzyme, a drop completely restorable by exogenous spermine [also see (400a)], perhaps results in a general decrease in production of messenger RNA. This fits the observation that orchiectomy results in a decreased ability of prostatic ribosomes to incorporate amino acids, whereas administration of testosterone rapidly restores the capacity for protein synthesis. Thus, it can be imagined that all of these phenomena turn on the control of polyamine synthesis by testosterone.

How are the polyamine levels increased by testosterone? Williams-Ashman has discussed one group of answers at the enzymatic level (457). The prostate lacks the ability to generate ornithine from arginine via arginase in the urea cycle or via glycine transamidinase to form creatine. However,

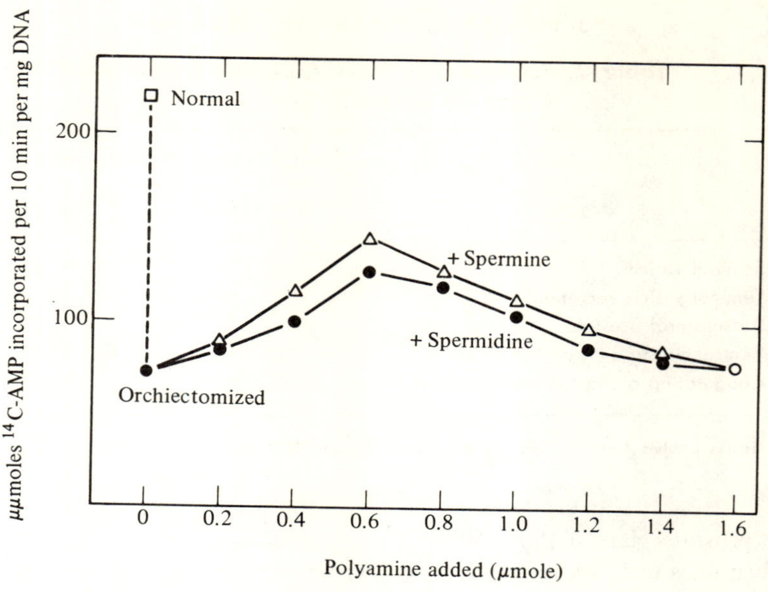

(a)

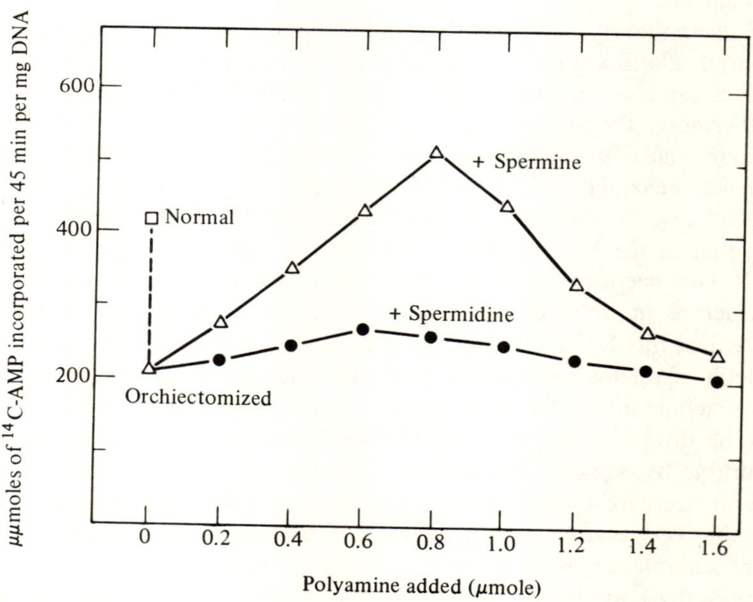

(b)

Fig. 38. (a) Dose-response curve for the effect of spermine and spermidine on DNA-dependent Mg^{++}-activated RNA polymerase activity in isolated prostate nuclei from orchiectomized rats (47). (b) The corresponding curve for the effect of the polyamines on DNA-dependent $Mn^{++}/(NH_4)_2SO_4$-activated RNA polymerase in comparable nuclei. The polyamines were added after 15 min of preincubation.

ornithine will arise in prostate from transamination to glutamic-γ-semialdehyde.

Ornithine decarboxylase is the next place to look; the enzyme has been purified by this group from prostate. The enzyme is soluble and is pyridoxal phosphate-dependent. Unlike the ornithine decarboxylase described by Raina and Jänne in regenerating rat liver (306), the prostatic enzyme was competitively inhibited by the natural polyamines, as presented in Fig. 39. The K_i's for putrescine, spermidine, and spermine are 1.2, 2.7,

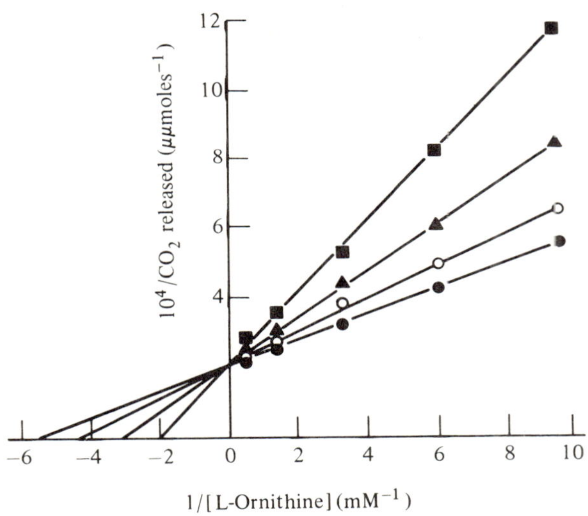

Fig. 39. Effect of putrescine, spermidine, and spermine on ornithine decarboxylase of rat prostate at various concentrations of ornithine (292). Results are shown for incubations with ornithine alone (●) and in the presence of 2 μmoles of spermine (O), of spermidine (▲), and of putrescine (■).

and 9.0 mM, respectively (292). Thus, spermine is almost non-inhibitory in this system.[5]

The level of enzyme in the gland falls about 80% after castration, but after administration of testosterone it increases to about 70% of normal (457). Once again we find this enzyme responding very early to hormonal stimuli.[6]

---

[5] Recent studies of Jänne and Williams-Ashman (188a) have indicated that the molecular weight and activity of the enzyme are markedly affected by dithiothreitol.

[6] According to a recent report, injection of estrogen into immature chicks results in a 30–50-fold increase in ornithine decarboxylase of oviduct within a 4-hr period. The enzyme is increased when isolated oviducts were incubated in vitro in the presence of estrogen; this appears to be the first known example of an in vitro enzyme response to estrogen (60a).

Pegg and Williams-Ashman then went on to study another prostatic enzyme, the decarboxylase of S-adenosylmethionine (291), the first enzyme of the sequence shown by the Tabors in *E. coli* that leads to spermidine synthesis. The prostatic enzyme was found to have several differences from the *E. coli* enzyme. In contrast to the bacterial enzyme, which has an absolute Mg^{++} requirement (454), the prostatic enzyme does not require this cation (293). Furthermore, the prostatic decarboxylase was found to be markedly and specifically stimulated by putrescine [Fig. 40(a)], unlike

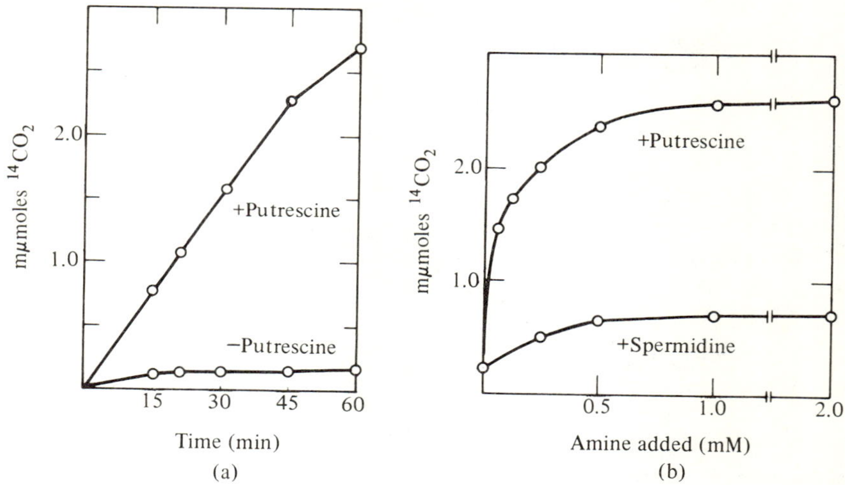

Fig. 40. (a) Time course of the decarboxylation of S-adenosylmethionine in the presence and absence of putrescine (293). (b) Effect of putrescine or spermidine on decarboxylation of S-adenosylmethionine.

the bacterial enzyme. At the degree of purification attained (293, 294), the decarboxylase was also inseparable from the aminopropyl transferase, whereas the two bacterial activities are easily separated. Inhibitor studies with the animal enzyme have suggested the presence of pyridoxal phosphate, which is absent from the *E. coli* enzyme.

In making this comparison at this time, we should realize that the soluble mammalian enzyme has only been purified 84-fold (293), in contrast to an approximately 800-fold purification to apparent homogeneity for the bacterial enzyme (454). Although the former degree of purification is not insignificant and numerous tests have failed to separate decarboxylase and aminopropyl transferase activities, some workers may prefer to reserve judgment of the apparent differences.[7] The recent extensive purification of the *E. coli* adenosylmethionine decarboxylase has permitted the definition of various properties of the enzyme against which the mammalian enzyme can be compared. Thus, the bacterial enzyme has a molecular weight of 113,000 with subunits of 15,000, appears to lack pyridoxal phosphate, and is not inhibited by isonicotinic hydrazide, but instead does contain covalently bound pyruvate. The presence of pyruvate appears to relate in this instance to inhibition by reagents reactive with carbonyl groups (293). The unusual property of the enzyme, bound pyruvate, has recently been described for the histidine decarboxylase of *Lactobacillus 30a* (319) and the proline reductase of *Clostridium sticklandii* (172) All of these enzymes similarly lack pyridoxal phosphate. The mammalian decarboxylase is inhibited by isonicotinic acid hydrazide (291, 293). Although the reversal of this inhibition by pyridoxal phosphate suggests that the latter is a cofactor in this system, the point warrants clarification.

The mammalian enzyme, which appears to have fused the entire sequence, is also activated by spermidine and produces spermine, thereby accounting for the formation of this polyamine. The effects of putrescine and spermidine on the decarboxylation reaction are shown in Fig. 40(b).

The stimulation of the decarboxylation by spermidine is not as great as by putrescine. Nevertheless, in both instances the reaction leads to aminopropyl transfer; that by putrescine, forming spermidine; and that by spermidine, forming isolable appropriately labeled spermine. These two activities of the mammalian decarboxylases cannot be separated, nor can they be distinguished by their thermal stabilities (294). Putrescine is a competitive inhibitor of spermine synthesis, which attains only 25% of the rate of spermidine synthesis at saturating levels of all reactants. S-adenosylethionine is decarboxylated in the presence of putrescine or spermidine at about 10% of the rate at which the methionine derivative reacts (294), unlike the complete lack of activity of the ethionine derivative with the *E. coli* enzyme. No evidence has been obtained in any system for a trans-

---

[7] Two groups have recently reported the isolation of S-adenosylmethionine decarboxylase lacking full activity as an aminopropyl transferase (305a, 188b). As described in the paper of Jänne and Williams-Ashman (188b), a very extensive purification produced a putrescine-stimulated decarboxylase which did not generate spermidine in the presence of putrescine or spermine in the prescence of spermidine. It is not at all clear that the activities are separate and uncoupled within eucaryotic cells. The K_m for decarboxylated S-adenosylmethionine in the aminopropyl transfer reaction appears to be too high (2.5×10^{-5}M) to permit this compound to serve efficiently as a free substrate in a system in which S-adenosylmethionine (K_m of 5×10^{-5}M) is decarboxylated in the presence of putrescine and generates spermidine at identical constant rates for both reactions (293). It is possible that the fractionation procedures inactivated sites or subunits of a single enzyme complex capable of coupled decarboxylation and polyamine synthesis.

ferase that takes 2 molecules of spermidine to 1 molecule of spermine plus 1 molecule of putrescine.

The prostatic enzyme is soluble, but a comparison of its molecular weight with the *E. coli* enzymes is not yet available. A knowledge of the structures of the bacterial and prostatic enzyme will be of great interest from the point of view of the acquisitions of function apparently represented by the fusion of activities, the positive effector activity, and the ability to synthesize spermine.

The intermediate, decarboxylated S-adenosylmethionine, seen in Fig. 15 (185, 186), is actually a better substrate than the parent compound in spermidine and spermine synthesis; this is shown in Fig. 41 (293, 294). Nevertheless, it is possible that this intermediate is not free in normal biosynthesis in the mammal. Do procaryotic cells and eucaryotic cells differ in this possibly important respect? The compound is a methyl donor in a number of transmethylations of histamine, acetylserotonin, and of homocysteine (467), but it appears unlikely that it is a major methyl donor in eucaryotic cells.

The possible fusion of several activities as well as the possible elimination of a key intermediate in the free state may reflect the evolution of a system that finds it useful to concentrate these functions and minimize problems of providing high concentrations of critical intermediates. If the main purpose of the products of such synthetic reactions is to facilitate accumulations of RNA, it can be imagined that the nucleolus itself, which is known to contain S-adenosylmethionine for methylation of ribosomal RNA,

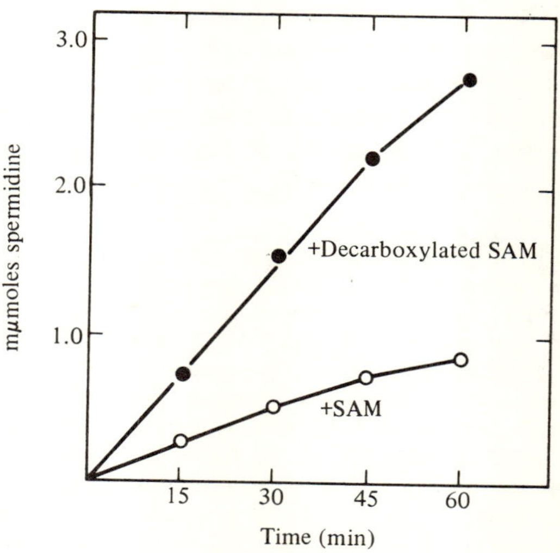

Fig. 41. Time course of formation of spermidine from putrescine and decarboxylated adenosylmethionine (SAM) or from adenosylmethionine (293).

might well be the organelle in which the structural elements essential to both RNA and polyamine syntheses have evolved. No direct evidence in support of this notion concerning the site of polyamine synthesis has yet been obtained, although interesting preliminary studies have been reported with an anucleolate mutant of the amphibian *Xenopus laevis* (334). The mutant makes far less ornithine decarboxylase and spermidine than do normal tadpoles at a comparable stage (339). Significantly, the adenosylmethionine decarboxylase was undetectable in similar stages of the anucleolar mutant (334).

A product of spermidine and spermine synthesis is 5-thiomethyladenosine (Fig. 15). This substance does not accumulate in the prostate (317), as it does in some microbial cultures. However, certain microorganisms are known to cleave thiomethyladenosine to adenine and methylthioribose (357), and so we wonder if this or other enzymes capable of metabolizing thiomethyladenosine are not fairly widespread.

The ventral prostate is much higher in the SAM decarboxylase than are other tissues (Table 3). After castration the level of the enzyme falls

Table 3. Spermidine Levels and Activity of SAM Decarboxylase in Male Accessory Organs of the Rat (457)

Organ	*Spermidine content (μmoles/g wet wt tissue)*	*SAM decarboxylase activity (mμmoles $^{14}CO_2$/ 30 min/mg protein)*
Coagulating gland	0.54	0.009
Seminal vesicle	1.09	0.031
Dorsolateral prostate	2.83	0.091
Ventral prostate	5.41	0.430

over 90% (457); but with the administration of testosterone, the enzyme makes an extraordinary recovery to higher than normal levels, in general simulating the pattern observed for ornithine decarboxylase, as shown in Fig. 42.

Conclusion

In many of the phenomena related to normal and pathological growth, including tumor development (335), the increase of the polyamines and of the enzymes essential to their synthesis are among the earliest reactions

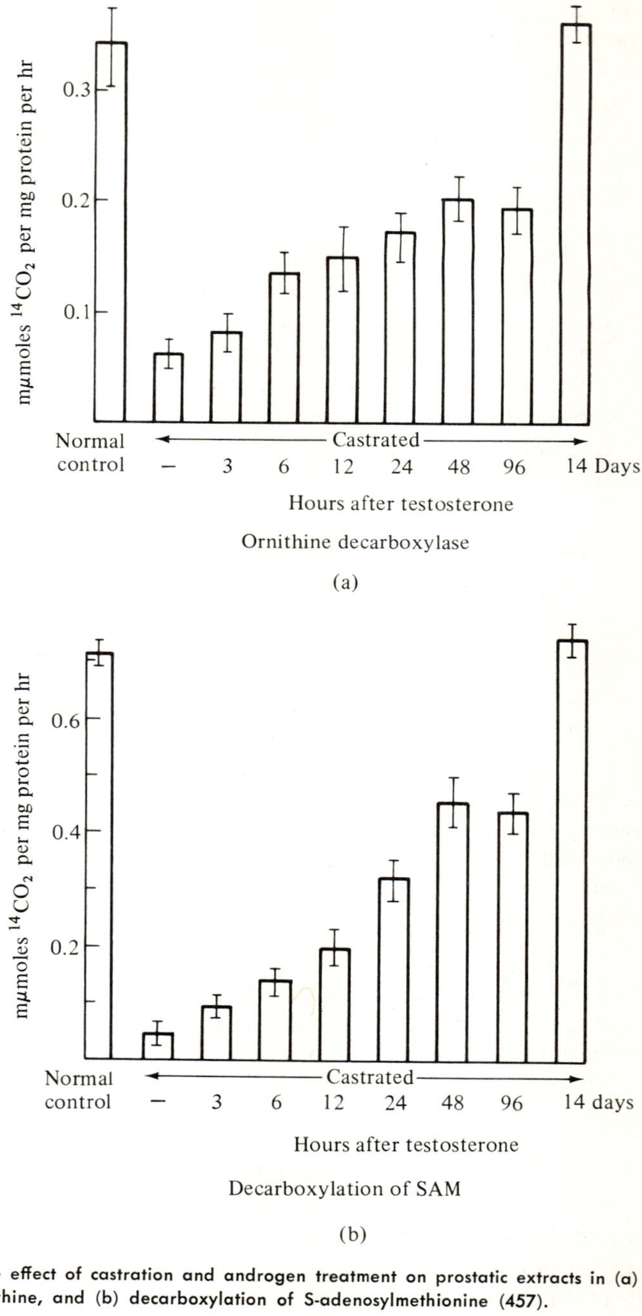

Fig. 42. The effect of castration and androgen treatment on prostatic extracts in (a) decarboxylation of ornithine, and (b) decarboxylation of S-adenosylmethionine (457).

that have been detected. How are these startling events effected? We do not know, but at least we can begin to ask questions relating to particular enzymatic activities. Furthermore, what exactly do the polyamines themselves do? Their formation precedes or is concomitant with RNA synthesis and, eventually, with protein synthesis. The correlation of polyamine synthesis and RNA synthesis is sufficiently tight in all systems studied to suggest a causal relation. Occasionally but not too often, the appearance of polyamines relates to DNA synthesis, but perhaps these events are less directly connected.

How shall we dissect the roles of the polyamines in cells? It seems appropriate at this time to turn to microorganisms whose manipulation for experimental purposes provides more controllable metabolic opportunities. The next chapter continues therefore, with the dissection that can be carried out in the study of procaryotic organisms.

Chapter 3

Bacterial Physiology and the Polyamines

"In exponentially growing cells, the aggregate flow of ornithine and arginine into the polyamine pool is approximately equal to the incorporation of arginine into protein."

RINALDINI AND MAAS (320)

In the preceding chapter, data were presented on the distribution of the polyamines in eucaryotic cells. We have discussed the biosynthetic origin of these compounds and the presumably important acquisition of an enzyme that can synthesize both spermidine and spermine in such cells. In certain systems, as in the study of Ham (145), these compounds have been found essential to the growth of animal cells in tissue culture.

The substances are present in millimolar concentrations in animal cells and bear a stoichiometric relation to the RNA content of the cells. For example, in the striking study of Seiler (352) the contents of RNA and spermidine were linearly related (Fig. 32) under conditions in which starvation reduced the RNA content of mouse liver, or substances such as phenobarbital caused liver enlargement and increase of liver RNA.

In following the time course of metabolic events during embryonic growth or in the response of the mammalian liver to nephrosis, hepatectomy or growth hormone, it was shown that the characteristic growth response was preceded by synthesis of RNA and of polyamines. In fact, the synthesis of the polyamines preceded that of RNA in many instances. The appearance of these organic cations was preceded by the extraordinary specific synthesis of the enzymes, ornithine decarboxylase and S-adenosylmethionine decarboxylase, which are essential for production of putrescine and spermidine respectively. These events have been described in detail for the response of the prostate to testosterone in the castrate rat. It was shown also that the polyamines stimulated the two disinct RNA polymerases of prostatic nuclei.

It is evident, therefore that a very good basis can be offered for the theory that the polyamines play essential roles in the various types of growth responses of eucaryotic cells and for their RNA metabolism. In this chapter, we examine the data on the biosynthesis and metabolism of the polyamines in bacteria and consider the data relating these compounds to the biosynthesis of essential polymers, mainly nucleic acids. In the course of this inquiry, we shall begin to formulate ideas concerning the molecular roles of these cations, a problem we explore in detail in the last chapter.

Studies on Bacteria

Current progress on the polyamines in microbial physiology derives from three major types of studies that appeared between 1945 and 1958. Snell and Herbst and the Israeli school demonstrated two roles of the polyamines in bacterial growth. The first study revealed true requirements for putrescine and spermidine in the growth of *Hemophilus parainfluenza* and some lactobacilli. Herbst subsequently developed this result to reveal a correlation between polyamines and the RNA content of the cells. The Israelis, studying more fastidious organisms, i.e., *Pasteurella, Mycoplasma,* and halophiles, discovered that substances present in complex media, which proved to be spermidine and spermine, also protected these cells from lysis.

In the second type of study, the systematic work of the Tabors and their collaborators on the biosynthesis of the polyamines developed analytic methods for these substances and used these to reveal the presence of putrescine and spermidine and their monoacetyl derivatives in bacteria. They then demonstrated unequivocally the in vivo path of conversion of putrescine to spermidine and elucidated this path at the enzymatic level. Their work related the metabolism of ornithine, arginine, and methionine to the biosynthesis of the polyamines.

Polyamines and Phage Multiplication

In 1957 a few molecular biologists were alerted to the existence of the polyamines when Hershey (156) discovered the existence of two ninhydrin-positive non-amino acid components in the T-even viruses. These compounds accounted for 1.5% of the carbon of these viruses. Hershey was unaware of the chemical identity of his substances and did not speculate on their possible relation to the known polyamines. Nevertheless, he devised a sound chromatographic method for their isolation, demonstrated their derivation from arginine, showed that they were injected into the host with virus DNA and that similar substances existed in the uninfected host. Hershey recognized that their origin in arginine probably conferred basic properties to the substances and suggested that they might neutralize charges on phage DNA. Ames and Dubin (8, 7) identified the compounds present in phage and went on to prove Hershey's hypothesis. They analyzed T4 phage and, in accounting for almost 80–90% of the total cation of the virus, showed that the polyamines accounted for about 40–50% of the identifiable cation. Their paper with this analysis (7), presented in Table 4, is one of the few papers among the many thousands on viruses and

Table 4. Cation Content of T4 Phage (7)

	Phage incubated in	
Cation	*$MgCl_2$*	*$CaCl_2$*
	meq/eq P	
Putrescine^{++}	250	290
Spermidine^{+++}	75	87
Ca^{++}	<2	96
Mg^{++}	340	360
Na^{+}	90	30
K^{+}	60	<30
Total meq per eq P	815	863

nucleic acids that attempts to account for the cations as well as the other constituents of these materials.

In the paper of Hershey (156), it was shown that the polyamines, i.e., the two low molecular weight substances derived from arginine, which he designated as substance A, were transmitted unchanged from parent to offspring. The substances were presumably associated with virus DNA. However, substance A, added to the medium, competed with the transmission of this material from parent to offspring. Noting also its separability from DNA and protein during dialysis of disrupted phage against buffer, Hershey concluded that these materials were probably neither genetically nor structurally important constituents of the phage DNA. It seems probable that Hershey was defining "importance" in a somewhat limited context, although he did not elaborate on this matter.

Arginine, added to cultures of infected bacteria, is converted to polyamine, subsequently isolable from the phage (156). Ames et al. showed that exogenous ^{14}C-putrescine is converted to ^{14}C-spermidine, isolable from phage with only moderate dilution of the latter (8). A considerable net synthesis of putrescine and spermidine does occur during infection, as presented in Fig. 43 (69). Immediately after infection small amounts of the compounds leak from the cell. Concomitantly with DNA synthesis, the substances are taken back into the cell and a net increment can be detected shortly thereafter. If these substances are made and packaged into phage, we may assume that their synthesis is in some degree important to phage multiplication.

Ames and Dubin went on to show that the cation content of phage could vary significantly, as shown in Table 4 (7), with the polyamines Ca^{++} and Mg^{++} making up at least ¾ of the detectable cation neutralizing the DNA. Almost all of the polyamines were displaced from the permeable O mutants of T4 by washing with buffer containing 0.01 M Mg^{++}. Although this experiment is often described as suggesting that polyamines can be

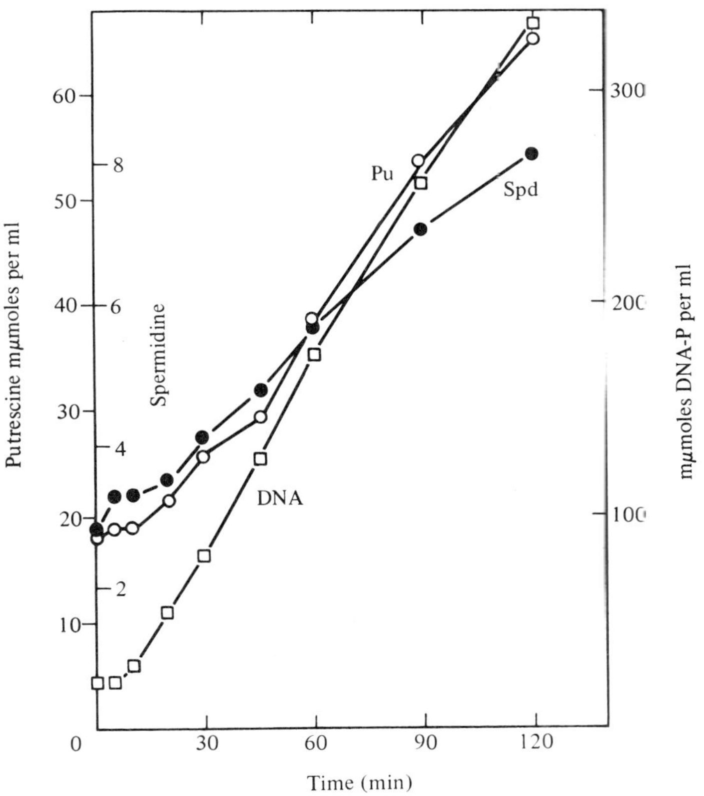

Fig. 43. The net synthesis of polyamines and DNA in T6r⁺-infected *E. coli* strain B (69).

largely displaced without loss of phage activity, the low infectivity of the control stock calls for additional data to prove this point unequivocally. In any case, it must be remembered that polyamine-free DNA is injected into a bacterium rich in polyamine.

Permeable phages, such as P22 grown on *Salmonella typhimurium* in a synthetic medium or T3 and T5, also lack polyamine. It may be that the polyamine initially present in these phages is lost during isolation in solutions containing various electrolytes.

When P22 is grown on *Salmonella typhimurium* containing spermine in a medium free of spermine (since spermine blocks adsorption and penetration of this phage), the phage can be isolated, containing small amounts of spermine. This polyamine is less readily dissociable from nucleic acids than are putrescine and spermidine. T4 can be grown in *E. coli* in minimal media containing spermine, and the spermidine of this phage will be almost totally replaced by spermine.

That the stability of many phages is improved by added Mg^{++} or Ca^{++} is a frequent observation, but few workers have similarly tested the polyamines. When divalent cations are removed by washing T5 phage with citrate or ethylenediamine tetraacetate (EDTA), the presence of low concentrations of spermidine or spermine prevents inactivation (416). Putrescine and cadaverine are also effective but at much higher concentrations. Ames and Dubin have suggested that the polyvalent cations, including the polyamines, are important in stabilizing the virus DNA.

We can infer from several types of experiments that the divalent cations or polyamines are essential to phage biosynthesis. The small DNA coliphage, ϕR, has a small burst size on *E. coli* strain C at low concentrations of Mg^{++}. The yield of virus may be markedly stimulated by higher concentrations of Mg^{++}, Ca^{++}, etc. among metallic divalent cations or by the polyamines. The effects of these cations occur mainly during the latter part of the latent period, i.e., the period of phage maturation (439).

Other experiments have involved the use of two different types of mutant phage. As is well known, the mutant strain T4rII is unable to multiply in *E. coli* K12 (λ) although it can multiply in *E. coli* B. Garen showed that high Mg^{++} permits multiplication of T4rII mutants in *E. coli* K12 (λ) when this cation is added at the eighth minute after infection (127). Ferroluzzi-Ames and Ames (108) then went on to show that spermidine alone will restore phage synthesis, as in Fig. 44. Noting that infection leads to leakage of polyamines, they added putrescine and spermidine to the medium at the concentration found in the bacterium and almost completely restored virus multiplication. However, they were unable to determine if these substances merely preserved the weakened membrane or facilitated DNA synthesis more directly.

In our laboratory, Dr. A. Dion has explored selected strains of bacteria

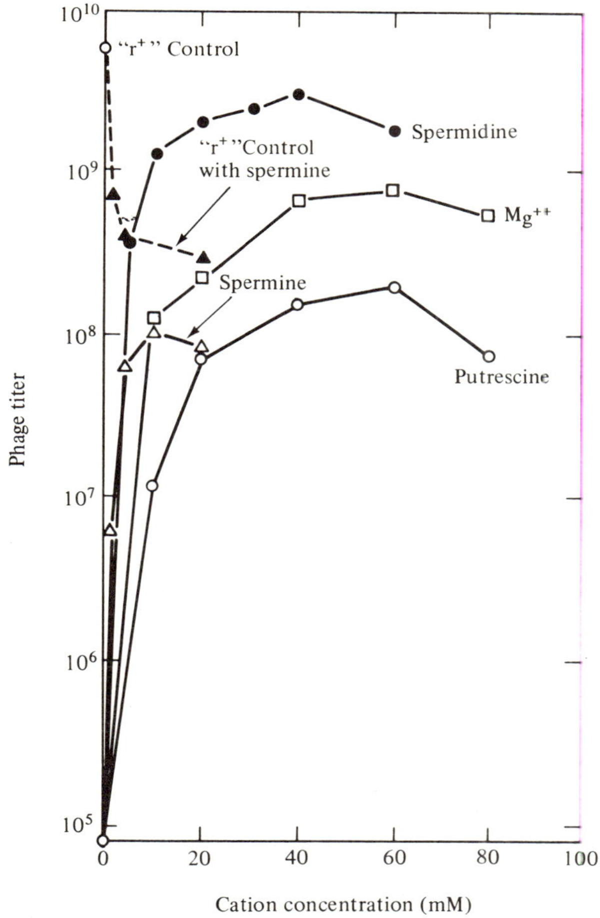

Fig. 44. The effects of spermidine, putrescine, and Mg^{++} in restoring T4rII multiplication in *E. coli* K12 (λ) (108). Each cation was added at the 6th min after infection.

infected by a phage mutant designated as amN116. Infection by this phage mutant is characterized by a marked delay in DNA synthesis in cells having a low concentration of Mg^{++}. It was suggested that this mutant is unable to direct the development of an essential Mg^{++} pump in EDTA-treated cells (143). As shown in Fig. 45, a comparison of DNA synthesis in the restrictive cell reveals that low concentrations of spermidine will help DNA to start earlier and to continue at a normal rate. Putrescine is ineffective in this activity. It may be mentioned that unlike the rII infection in which putrescine is lost from the cell (108), infection by amN116 produces an accumulation of this diamine to greater than normal levels.

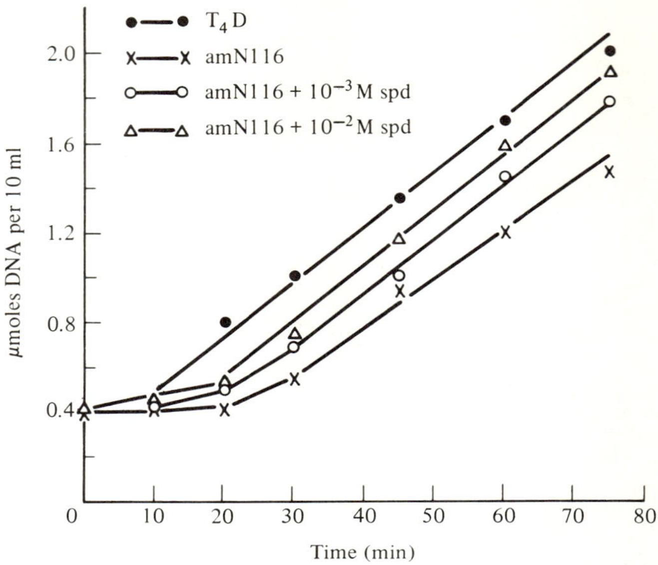

Fig. 45. The effect of spermidine on the inception of DNA synthesis in infection by the T4 phage mutant, amN116. (A. Dion and S. S. Cohen, unpublished data.)

It is clear, therefore, that in phage multiplication spermidine or Mg^{++} can fill significant functions related to the synthesis and packaging of phage, but it is not yet evident what these functions may be.

High concentrations of putrescine (100 mM) markedly inhibit DNA synthesis (355). It has been found (73a) that about 50% of the few T-even phage progeny that are formed under such conditions are small-headed forms. Spermine, spermidine, and agmatine have no apparent effect on head formation (73a).

Thus, polyamines are present in phage and neutralize much of the virus DNA. They are mainly synthesized along with virus DNA. In infected cells from which polyamine has leaked and that lack the ability to concentrate cations, exogenous Mg^{++} or the polyamines (particularly spermidine) restore the ability to make DNA and virus. Furthermore, an excess of putrescine inhibits DNA synthesis and distorts the assembly of intact phage. These data, which are strongly suggestive of a crucial if not essential role of the polyamines in virus multiplication, must be further clarified.

Although more detailed answers will be seen later in studies of the relation of the polyamines to the synthesis of RNA, the following questions are posed: (1) is the apparent concomitance of production of polyamine and nucleic acid real or can the two types of syntheses be completely separated, and (2) if the two synthetic systems are related, does nucleic acid synthesis determine polyamine synthesis or does polyamine synthesis effect controls on nucleic acid synthesis? To answer these questions for the relation of the polyamines to the synthesis of phage DNA, we set up systems in which (1) damaged or mutant phage are used to infect under conditions in which DNA synthesis is prevented, delayed, or inhibited; and (2) mutant bacteria low in putrescine, and apparently capable of only very low levels of putrescine biosynthesis, are infected by normal phage.

In systems in which the synthesis of virus DNA was prevented or inhibited, it was found that polyamines continued to be synthesized under conditions in which little or no DNA was made. These results are presented in Fig. 46.

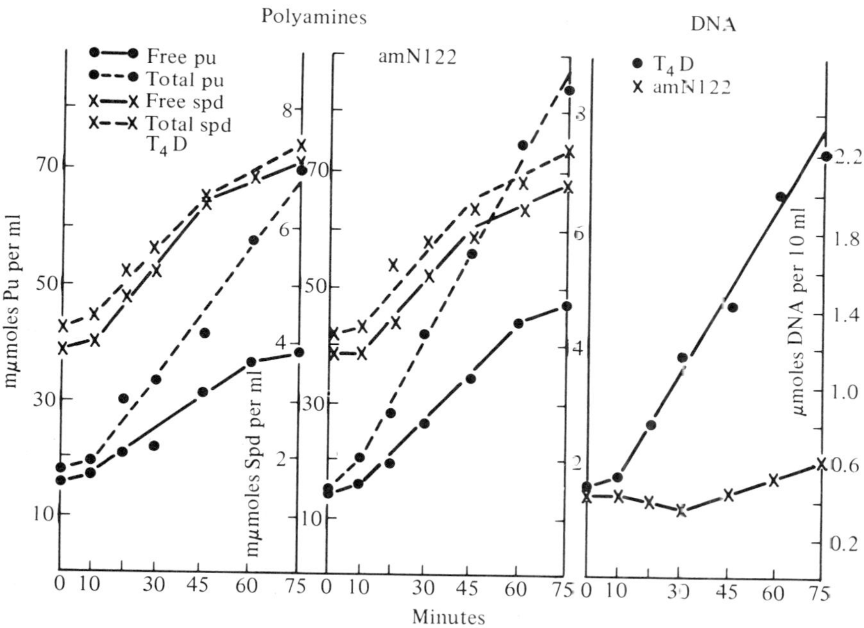

Fig. 46. The synthesis of polyamines and DNA in cells infected by wildtype $T4r^+$ phage and amber mutant $T4r^+$-N122. (A. Dion and S. S. Cohen, unpublished data.)

Preliminary results exploring the second situation have indicated that the beginning and rate of DNA synthesis is largely dependent on the presence of exogenous polyamine. These results, shown in Fig. 47, appear to affirm the essential role of polyamines in the synthesis of T-even phage DNA in polyamine-deficient mutants, since exogenous polyamine is necessary to achieve the normal synthesis of DNA in this system. A significant de novo synthesis of spermidine, derived from a low level of putrescine synthesis in the infected mutant, may account for the relatively late DNA synthesis in the depleted cell. Moreover, some turnover of ribosomal RNA

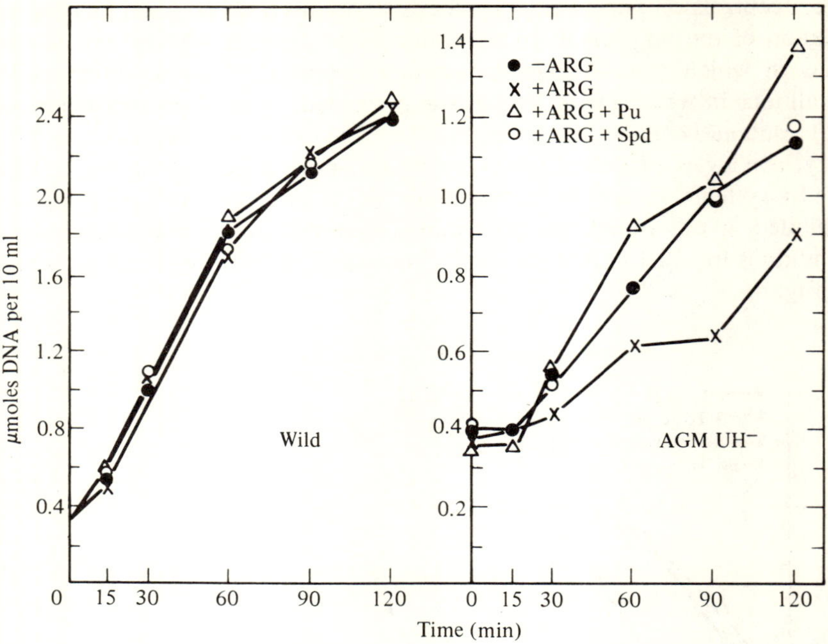

Fig. 47. The synthesis of DNA in putrescine-deficient T4D-infected *E. coli* strain K12 in various media. (A. Dion and S. S. Cohen, unpublished data.)

in such a system can be expected to release spermidine for internal re-utilization. These experiments have been performed with the bacterial mutants supplied by Dr. W. Maas and are continuing.

As observed earlier, for the most part the polyamines relate mainly to RNA synthesis and few systems are known in which these substances can be shown to effect the synthesis of DNA. The isolated nuclei of *Physarum polycephalum* is one instance in which the incorporation of a deoxyribonucleoside triphosphate is markedly stimulated by the addition of spermine (41). This system has not been analyzed further. The newly available polyamine-deficient bacterial mutants infected by phage appear to present another major opportunity to study the control of DNA synthesis by these cations. As is seen later, the mutants described by Maas (169) to date are fairly complex and somewhat leaky; the production of a polyamine deficiency in these organisms at present depends on a technique that is time-consuming and requires improvement. In short, we can look forward to the production of better mutants for the performance of this type of experiment.

Infection with ultraviolet-irradiated T3 phage induces the production of an enzyme that destroys S-adenosylmethionine (63, 129). This system may well provide approaches to explore T-even phage multiplication in cells relatively incapable of synthesis of putrescine or spermidine. In this system, however, incompleteness of the metabolic blocks or the release of host polyamine from nucleic acid-binding sites may confuse the interpretation of the results.

Polyamines and the Bacterial Chromosome

Virtually nothing is known of the possible in vivo association of the polyamines with bacterial DNA. Johnson and Bach (196) have reported that radioactive spermine taken up in a non-exchangeable form by spheroplasts of *E. coli* is associated with both the DNA and RNA isolable from the cells. They have interpreted (257) this result as helping to explain an antimutagenic effect of polyamines, an effect originally reported by Sevag (354). In confirming and extending (194, 195) the report of Sevag, they have shown that the antimutagenic action requires the continuous presence of spermine during apparently normal growth. The response of cells to many types of mutagenic agents seems to be diminished by the presence of spermine. Nevertheless, the mechanism of the antimutagenic action remains unexplained. Nor does the experiment of Johnson and Bach (196) prove the presence of normal polyamines on the DNA of untreated growing cells.

Michaels and Tchen (257) have reported that the polyamine content of nucleated and enucleated ("minicell") *E. coli* were approximately the same when calculated on the basis of RNA content. I do not believe that these data exclude the possible presence of spermidine on bacterial DNA. However, no one has attempted to isolate bacterial DNA under conditions that might permit the detection of bound spermidine. Even if this were done, if bound polyamine were found, it would have to be proved that the bound polyamine did not associate with the DNA after cell disruption.

Polish workers (216) have reported that 2 mM spermine increased zygotic induction of Hfrλ lysogens but reduced the frequency of recombination between mated Hfr and F^- bacteria. In similar circumstances, 3 mM spermidine had no effect.

Most recently, Inouye and Pardee (179) have observed a marked effect of the polyamines on the course of bacterial division. When certain mutant strains are starved for an hour for an amino acid and the amino acid is restored, cell multiplication is synchronized. This result is true for an arginine-requiring cell blocked before ornithine, but not so for cells blocked after ornithine, as can be seen in Fig. 48. In the first instance, putrescine (which is made in excess in an organism blocked between ornithine and arginine) prevents synchronization (Fig. 49). Spermidine,

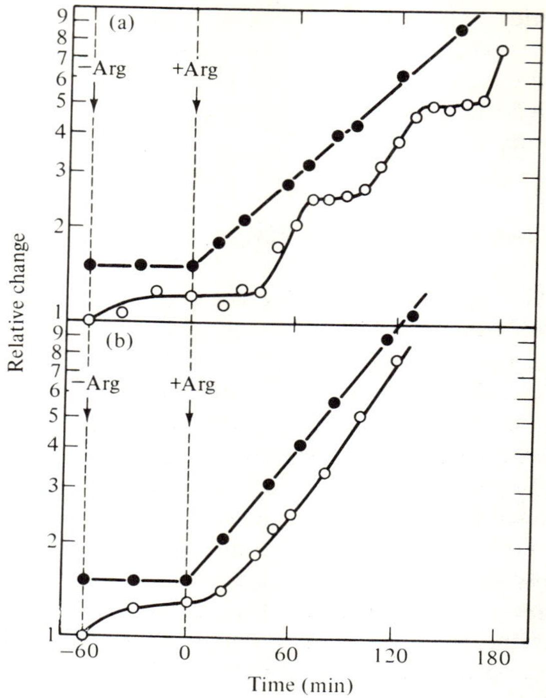

Fig. 48. Effects of arginine starvation on cell division (179). Arginine was added to cultures starved for 1 hr for the amino acid and optical densities (●) and total cell counts (O) were measured. (a) *E. coli* 15 T⁻arg-1⁻; (b) *E. coli* 15 T⁻ arg-2⁻.

added to the putrescine [in Fig. 49], restores synchrony. Furthermore, spermidine added to the second asynchronous organism converts its multiplication to a synchronous event. These authors have concluded that the polyamines are essential in permitting cell division and that this is in fact controlled by the molar ratio of putrescine to spermidine. The significance of this ratio in affecting RNA synthesis will be noted later.

Polyamines and Nutritional Deficiency

We have already mentioned several organisms requiring polyamines for growth. Among these we find an organism such as *Lactobacillus casei,* which was reported to lack any polyamine initially (154). In fact, without one of the polyamines in the medium this cell grows very slowly, but it sharply increases its rate of growth in the presence of these substances (204). Moreover, assay of the organism grown on a defined synthetic medium lacking polyamine has shown that *L. casei* accumulates considerable spermidine during the lag phase and in the course of division reduces

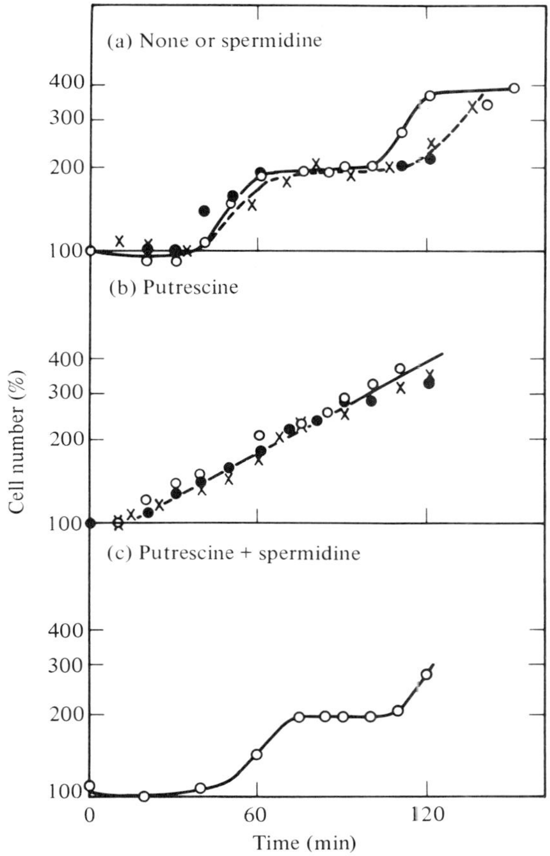

Fig. 49. Effects of polyamines on cell division of *E. coli* 15 T⁻arg-1⁻ (179). As in Fig. 48, arginine was added to a culture starved for 1 hr. (a) Total cell numbers, relative to initial values, without addition (●), plus 0.2 mM spermidine (O), viable cell counts in presence of 0.2 mM spermidine (x). (b) Total cell counts in presence of 0.12 mM (●) and 3.1 mM putrescine (O); viable cell counts in presence of 3.1 mM putrescine (x). (c) Total cell counts in the presence of both 3.1 mM putrescine and 2.0 mM spermidine.

this amount to a very low level at the time when the stationary phase is reached (80). Organisms analyzed for polyamine after overnight growth would be expected to have a low level of these substances. Actually, the growth response of *L. casei* to polyamine is so pronounced that Kihara and Snell (204) have asked if these substances were not the main growth stimulants in the "strepogenin" preparations of Wooley (391). Snell has also shown that the growth of many species of lactobacilli is stimulated by spermine or spermidine in the medium. A homolog inactive in stimulating growth, N-(3-aminopropyl)-1,6-hexanediamine, is a competitive inhibitor of spermidine or spermine in eliciting the growth response (141). Two roles have been suggested for the polyamines; one relatively non-specific, possibly in neutralizing and stabilizing labile polyanions, and a second that requires small amounts for a specific structural role. Chapter 4 considers how the second role may involve a small variation in the first type of binding.

Gram-positive organisms such as *S. aureus* appear to be more sensitive than Gram-negative *E. coli* to polyamines and their analogs, which have been studied not only for effects on growth but also for effects on RNA and protein synthesis (25). Some of these compounds have been observed to stimulate RNA synthesis in growing (25) or stationary (305) cultures, an effect whose significance will be discussed later in greater detail.

Mention was made in an earlier chapter that Kim (206) had found an unusual *Pseudomonas* species that appears to contain putrescine and to lack spermidine. This report was confirmed by Rosano and Hurwitz. However, they reported that a new polyamine, 2-hydroxyputrescine, is found in this strain (328). These workers had reported earlier (175) that *E. coli* grown in limiting Mg^{++} synthesized more spermidine; thus, ribosome-bound spermidine, but not putrescine, was found to vary inversely with ribosome-bound Mg^{++}. Unlike *E. coli,* in *Pseudomonas* (Kim) both ribosome-bound putrescine and ribosome-bound hydroxyputrescine (Table 5) vary inversely

Table 5. Effect of Extracellular Mg^{2+} on Ribosome-Bound Mg^{2+} and Putrescine (78)

Concentration of Mg^{2+} in medium	μmoles per μmole RNA-ribose			
	Mg^{2+}	Putrescine	Spermidine	Hydroxy-putrescine
Pseudomonas (Kim)				
4×10^{-5}M	0.075 ± 0.011	0.145 ± 0.020	0	0.038 ± 0.002
4×10^{-4}M	0.130 ± 0.010	0.071 ± 0.008	0	0.022 ± 0.003
4×10^{-3}M	0.240 ± 0.030	0.024 ± 0.003	0	0.013 ± 0.001
E. coli ML35				
4×10^{-5}M	0.090 ± 0.010	0.071 ± 0.066	0.027 ± 0.004	0
4×10^{-4}M	0.138 ± 0.015	0.059 ± 0.008	0.014 ± 0.002	0
4×10^{-3}M	0.240 ± 0.025	0.063 ± 0.009	0.007 ± 0.002	0
Pseudomonas fluorescens				
4×10^{-4}M	0.147	0.070	0.018	0

with ribosome-bound Mg^{++} (328). It has been postulated that in the latter organism, hydroxyputrescine fulfills the role or roles of spermidine. Obviously, the nature of these roles should be clarified in both organisms.

The Biosynthesis of the Polyamines in Some Bacteria

E. coli strains usually contain several times more putrescine than spermidine (421), although the ratios may vary from strain to strain grown in comparable conditions (309). On a wet weight basis, putrescine may attain an average concentration of 10–20 mM as compared to 1–2 mM for spermidine. As we shall see later, relaxed strains in exponential growth contain similar but slightly higher molar concentrations of these compounds (65). In the presence of methionine, spermidine production in several strains is markedly increased as is RNA synthesis. In relaxed cells, both spermidine and RNA accumulate, resulting in a marked inhibition of putrescine synthesis. A thorough study of the structure, kinetics, and inhibition of the purified ornithine decarboxylase of *E. coli* has not yet been carried out.

Exploration of putrescine biosynthesis in *E. coli* has revealed a complex network of both constitutive and inducible pathways. Figure 50 shows the two basic routes that are known in this organism, one via the decarboxylation of ornithine and a second via the decarboxylation of arginine to agmatine. In the latter, the subsequent elimination of urea via agmatine

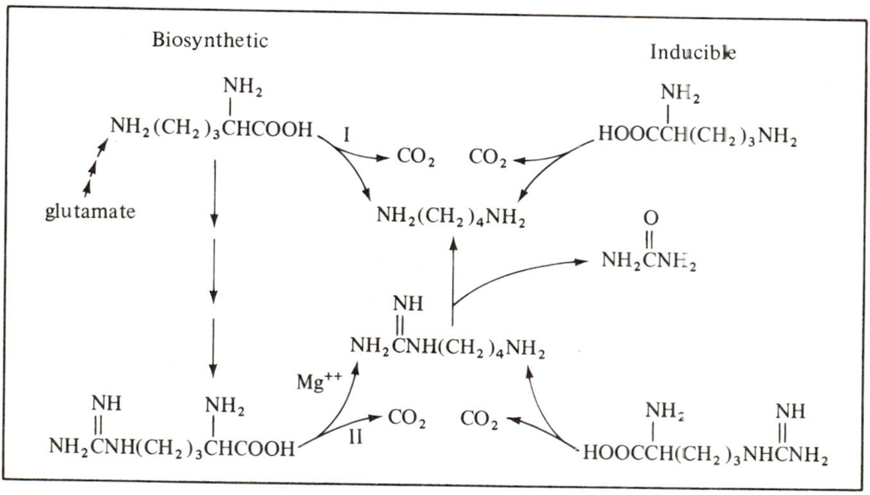

Fig. 50. Biosynthetic and inducible pathways of putrescine synthesis.

ureohydrolase produces putrescine. It should be clear, therefore, that urea production (*E. coli* lacks urease) will be a measure of putrescine biosynthesis via arginine.

A third route, given in Fig. 12, shows the formation of carbamylputrescine from agmatine and hydrolysis of the former to putrescine. This route operates actively in plants but is not yet known in bacteria.

The balance of the two pathways is modulated by a variety of factors. In many strains, exogenous arginine provides almost all of the putrescine, as measured by formation of both polyamine and urea (Fig. 51) (266, 269, 267). In such organisms, arginine represses ornithine biosynthesis and the putrescine is derived from agmatine, although the total cellular polyamine produced varies very little (267). In the strains presented in Table 6, in

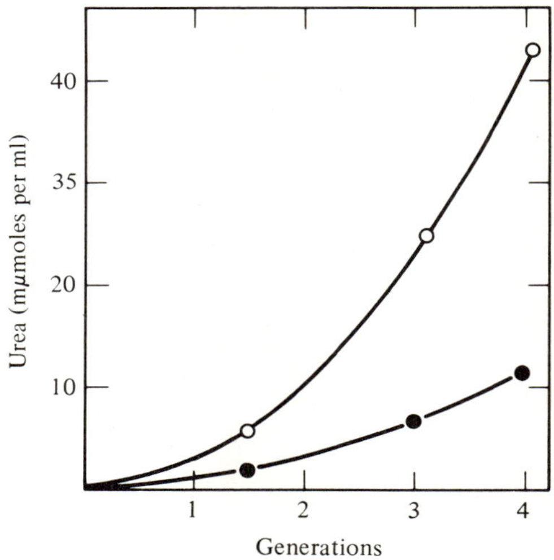

Fig. 51. Time course of urea formation by the non-arginine-requiring *E. coli* K12, strain 3000. Symbols: ●, unsupplemented medium; O, arginine-supplemented. Cultures were started at an initial density of 10^8 cells per ml and grew with a generation time of 1 hr (266).

the absence of arginine the bulk of the polyamine needed for growth is derived from ornithine (267). However, this phenomenon varies with the strain, e.g., *E. coli* AC 2514 vs. 3000, and applies to cells fairly low in ornithine transcarbamylase. Strains high in this constitutive arginine-synthesizing enzyme will make a higher proportion of putrescine via agmatine and urea production. Citrulline and arginine in the medium permit synthesis of putrescine through agmatine to a much greater extent than does exogenous ornithine in strain 30 SO-arg 4.

Thus, strains grown in the absence of arginine, like strain 3000, may also have a constitutive arginine decarboxylase (269). This enzyme has

Table 6. Effect of Culture Conditions on Urea Production (267)

Conditions	*Organism*	k_{urea}	k_{urea}/k_{put}
		mμmoles doublings⁻¹ (10⁹ cells)⁻¹	
Minimal	AC2514-1	1	0.02
+ Arginine	AC2514-1	56	0.98
Minimal	3000	12	0.23
+ Arginine	3000	46	0.87
Minimal			
+ Ornithine	30SO-arg 4	13	0.25
+ Arginine	30SO-arg 4	55	1.04
+ Citrulline	30SO-arg 4	36	0.68
+ Arginine, citrulline	30SO-arg 4	46	0.87
+ Ornithine, arginine	30SO-arg 4	32	0.60
+ Ornithine, citrulline	30SO-arg 4	30	0.57

a Mg^{++} requirement and a pH optimum of 8. It may be separated from the inducible enzyme that does not require Mg^{++} and has a pH optimum of 5.2.

The inducible enzyme produced during amino acid catabolism has been purified to homogeneity (33), crystallized, and shown to be a particle with a molecular weight of 850,000, comprised of 5 subunits (35, 36). The constitutive enzyme, apparently more actively involved in polyamine biosynthesis in the absence of exogenous arginine, has also been purified recently (208). Both enzymes require pyridoxal phosphate. The inducible enzyme is poorly inhibited by putrescine. Ornithine and spermidine are competitive with arginine; agmatine has a lower K_i (33). The constitutive enzyme is also inhibited by these polyamines, in a manner competitive partially with arginine and partially with Mg^{++} (460).

Both constitutive biosynthetic and inducible degradative ornithine decarboxylases are known in *E. coli* (126, 268). As in the case of the arginine decarboxylases, the constitutive ornithine decarboxylase has a much higher pH optimum than the inducible enzyme. The latter is far more heat labile (268).

The Tabors have extended this analysis to the growth of an ornithine auxotroph in a chemostat. Even under severe restriction of ornithine or arginine, about 5–18% of the amino acid is directed to formation of putrescine and spermidine (423). The polyamine content of such cells may equal about 25–30% of the arginine in the cell protein. The content of spermidine in the cells remains almost constant under all conditions, whereas putrescine can be extensively reduced. The Tabors suggested that the preservation of the polyamines even under restrictive conditions supports the idea that polyamines are necessary for the growth and survival of the organism.

When putrescine or spermidine is added to the chemostat, there is a marked decrease in biosynthesis of the polyamines from carboxyl-labeled ornithine (423). There is a slight initial reduction in liberation of isotopic CO_2 and a rapid fall to low levels of CO_2, as seen in Fig. 52. This is fol-

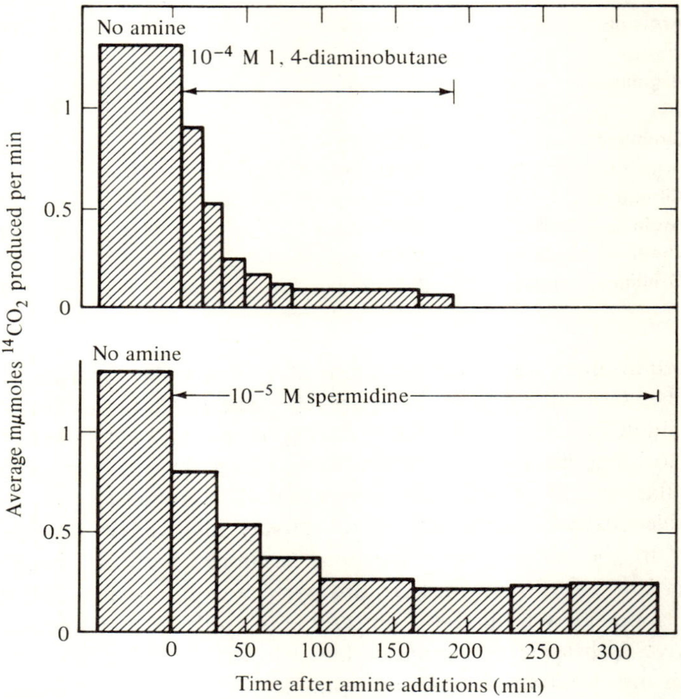

Fig. 52. Inhibitory effect of added 1,4-diaminobutane and of spermidine on the decarboxylation of carboxyl-labeled ornithine by *E. coli* 160-37 orn⁻ growing in a chemostat (423).

lowed by a repression of synthesis of both ornithine decarboxylase and arginine decarboxylase, as described in Table 7. Rinaldini and Maas have noted, however, that although exogenous putrescine and spermidine repress steps in polyamine synthesis, they do not repress or inhibit steps in arginine biosynthesis (320).

In a chemostat it can be shown that exogenous arginine does not mix completely with the endogenous pool in generating polyamine (424). It is possible that this fact also reflects a spatial separation of the two types of arginine decarboxylase.

Table 7. Ornithine and Arginine Decarboxylase in Extracts of *Escherichia coli* Orn⁻ Grown in Chemostat with and without Added 1,4-Diaminobutane or Spermidine (423)

Collection period	*Specific activity of ornithine decarboxylase*	*Specific activity of arginine decarboxylase*
Experiment I		
Before amine	0.45	0.089
After addition of 10^{-4} M 1, 4-diaminobutane		
0–1 hr	0.40	0.084
1–2 hr	0.30	0.037
2–3 hr	0.21	0.036
3–4 hr	0.15	0.040
4–5 hr	0.17	0.023
Experiment II		
Before amine	0.61	0.064
After addition of 10^{-4} M spermidine		
0–1 hr	0.48	0.072
1–2 hr	0.44	0.048
2–4 hr	0.38	0.034
4–5 hr	0.30	0.022
5–6 hr	0.32	0.027

Certain strains of *E. coli, Aerobacter aerogenes*, and *Serratia marcescens* possess an amine transaminase capable of transferring the amino group to α-ketoglutarate (256). This enzyme, which requires pyridoxal phosphate, is active on putrescine and cadaverine but not on spermidine and spermine (462). *E. coli* strain B slowly degrades ^{14}C-putrescine by this route, whereas many strains of *E. coli,* e.g., strain 15 TAU, do not appear to have this enzyme constitutively. Our laboratory strains 15, K12, B, etc. do not degrade ^{14}C-spermidine.

Kim has isolated a strain of *E. coli* strain B by growth on putrescine as sole nitrogen source; this organism contained a very active constitutive aminotransferase (205). During growth on NH_4^+, this organism maintained its putrescine and spermidine content. However, under conditions of nitrogen starvation, the cells reduced their putrescine content about 90% with only a 50% reduction in spermidine (205).

Putrescine-Deficient Mutants of *E. coli*

The existence of two pathways for putrescine biosynthesis, each of which is supported by both constitutive and inducible routes (Fig. 50), points to the essentiality of this key product. However, this very fact has made difficult the isolation of putrescine-deficient mutants. Several such mutant strains have only recently been described by two groups.

The first isolation, by Maas and collaborators (169), was a fortunate accident. Maas sought mutants for arginyl-tRNA synthetase by plating in minimal medium containing canavanine (168). One such canavanine-resistant mutant grew poorly in the presence of arginine, which represses the ornithine pathway. The organism grew significantly better in putrescine or spermidine but not in cadaverine, as seen in Fig. 53. We have shown that these cells are very low in putrescine and do not appear to convert spermidine to putrescine; these results would imply therefore the essentiality of spermidine. Table 8 (169) contains enzyme analyses that have

Table 8. Enzyme Levels in a Putrescine Auxotroph (169)

Enzyme	Product measured	Amt (μmoles) per min per mg of protein: Parent MA 98	Amt (μmoles) per min per mg of protein: Mutant MA 131	Per cent in mutant
Agmatine ureohydrolase	Urea	0.025	0.001	4
Arginine decarboxylase (constitutive)	CO_2	0.0185	0.0171	93
Arginine decarboxylase (inducible)	CO_2	0.78	0.28	36

demonstrated the relative lack of agmatine ureohydrolase in this organism. Maas' laboratory has isolated many more mutants deficient in either this enzyme or in the arginine decarboxylase (235).

Morris and Jorstad have also obtained similarly deficient organisms by examining the ability of mutagenized clones to produce urea (235). Two cultures, of about a thousand tested, had very low levels of putrescine when grown in arginine. One, CJ556-2, which had 4% of the normal arginine decarboxylase and a normal level of agmatine ureohydrolase, contained about 1% of the normal putrescine level. Under these conditions, the spermidine content of this cell did increase during growth (Table 9), indicating the existence of a leak in putrescine synthesis. The growth rate

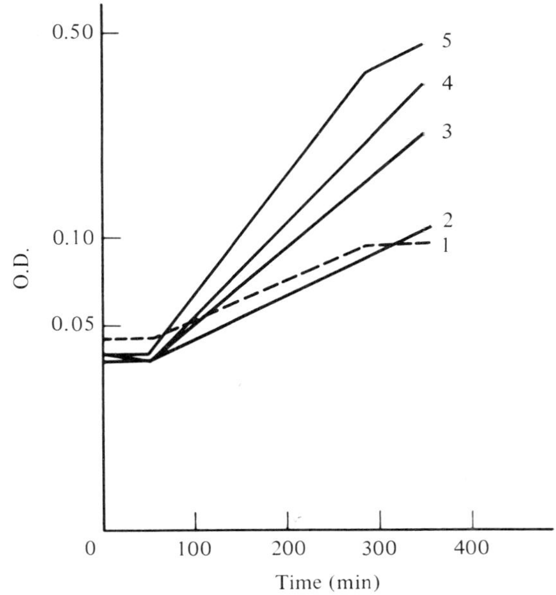

Fig. 53. Growth response of *E. coli* mutant MA 138 to arginine and to various polyamines in the presence of arginine (169). Curve 1, + arginine; 2, + arginine + cadaverine; 3, + arginine + spermine; 4, + arginine + spermidine; 5, + arginine + putrescine.

Table 9. Polyamine Levels in Mutants (235)

Strain	Growth condition	Polyamine levels [a]		
		Putrescine	*Spermidine*	*Agmatine*
BR9a	—arg [b]	22.1	3.8	<0.3
BR9a	+arg	26.4	4.0	<0.3
CJ556-2	+orn	25.6	3.7	<0.3
CJ556-2	+arg	<0.3	5.5	<0.3
CJ866	—arg	24.8	1.7	4.8
CJ866	+arg	5.0	2.2	20.2

[a] Expressed as mμmoles per 10^9 cells.
[b] Without arginine, —arg; with arginine, +arg; with ornithine, +orn.

of this organism in putrescine-deficient conditions was decreased only 10% (265).

Several of these mutants have been mapped, and the genes have been found to cluster near the gene, serine A. This observation suggests the possible existence of an operon for putrescine biosynthesis (168). No mutant is yet known for other steps in the pathways, ornithine decarboxylase, SAM decarboxylase, and the aminopropyl transferase. Obviously, the distribution of the genes of the polyamine pathways will be of great interest.

On Spermidine Biosynthesis and Metabolism

In a classical paper, Tabor, Rosenthal, and Tabor described the incorporation of isotopic putrescine into spermidine in bacteria and into spermidine and spermine in fungi (420). They also outlined the enzymatic steps of spermidine biosynthesis starting from methionine, ATP, and ^{14}C-putrescine and extracts of *E. coli*. In addition, the chromatographic separations of the numerous compounds, polyamines, and adenosylmethionine were described. Recently *E. coli* has been reported to excrete significant amounts of thiomethylribose, presumably releasing adenine from the thiomethyladenosine produced in spermidine biosynthesis (357, 350).

Very little work has been done on the intermediate aminopropyl donor, S-adenosyl 5′-S-methyl mercaptopropylamine, although its synthesis has been reported by Jamieson (186). The level of the compound attained in the cell, its penetrability, and utilizability have not been studied. No analogs have been made, despite the obvious interest of the compound. Although the enzyme that combines SAM decarboxylase and aminopropyl transfer in the animal cell can handle this intermediate (see footnote 7 in Chapter 2), there is no evidence that the compound exists in the free state in animal cells. We can ask once again whether this unusual difference in the biosynthesis of spermidine in bacteria and animal cells may not contribute to important differences between procaryotic and eucaryotic cells in general. We can also wonder if this synthesis could be restricted to nucleoli in animal cells. This cytological structure, absent in procaryotic cells, does concentrate reactions of ribosomal RNA synthesis and modification involving adenosylmethionine.

It was noted earlier that S-adenosylethionine is not decarboxylated by the *E. coli* enzyme (294) and is presumably not a spermidine precursor in this organism. However, ethionine is incorporated into protein and stimulates DNA synthesis in methionine-deficient *E. coli* strains. Nevertheless, it does not contribute ethyl groups to RNA and total RNA synthesis is not stimulated by ethionine in methionine deficiency (377).

A Spermidine-Glutathione Complex

Dubin (90) was the first to describe the presence in *E. coli* of a most unusual "spermidine-glutathione" derivative, amounting to no more than

10–15% of the total spermidine. The complex contained equimolecular amounts of cysteine, glycine, glutamate, and spermidine. The spermidine was not bound to the glutamate (414). We do know that in the presence of ATP, Mg^{++}, and glutathione, an enzyme of *E. coli* (422) converts spermidine to a derivative similar to that reported by Dubin. At the present time, however, nothing is known of the role of this compound or of its distribution in other organisms.[8]

Transport of Polyamines

The uptake of exogenous polyamines has been studied in *E. coli* (412). Acetylation of the compounds was minimized by addition of arsenite. Two processes were detected. At 0°, there is a rapid adsorption of radioactive spermidine and spermine, which can be released by a wash with non-radioactive polyamine or salt-containing media or by butanol or toluene. Very little putrescine is taken up at 0°. A similar binding of spermidine by cell walls of *B. subtilis* has been reported earlier (12).

A second process, which does not occur at 0°, is tied to metabolism. The accumulation of putrescine, as well as spermidine and spermine, is temperature-sensitive and is decreased by the omission of glucose, as shown in Fig. 54, or by the addition of dinitrophenol. The accumulation in a nitrogen-deficient medium that does not support growth may attain estimated internal concentrations of 30 mM for putrescine, 10 mM for spermidine, and 1 mM for spermine. The affinities of the polyamines for polymeric anions, however, make it difficult to relate these contents to true concentrations as dissolved entities. Amines accumulated by this transport process cannot be removed by washing the bacteria with solutions of the non-radioactive amines, although they are released by toluene. Intracellular ^{14}C-spermidine does not exchange with ^{12}C-spermidine or spermine added to the medium. Furthermore, low levels of exogenous spermidine do not exchange with accumulated putrescine, nor does the presence of spermidine interfere with the uptake of putrescine. However, spermidine does affect spermine accumulation. There thus appear to be at least two independent transport mechanisms, one for putrescine and another for spermidine and spermine. The existence in Gram-negative organisms of transport mecha-

[8] The Tabors have recently described (412a) the almost total conversion of intrabacterial spermidine to the glutathione complex as the bacteria become anaerobic in a slightly acidic carbohydrate medium, i.e., as the bacteria enter a stationary phase. This complex is hydrolyzed to free spermidine by a newly discovered peptidase as the bacteria in stationary phase are diluted and permitted to enter a logarithmic phase of growth. The possible relations of these phenomena to the origins of these developmental changes have not yet been clarified.

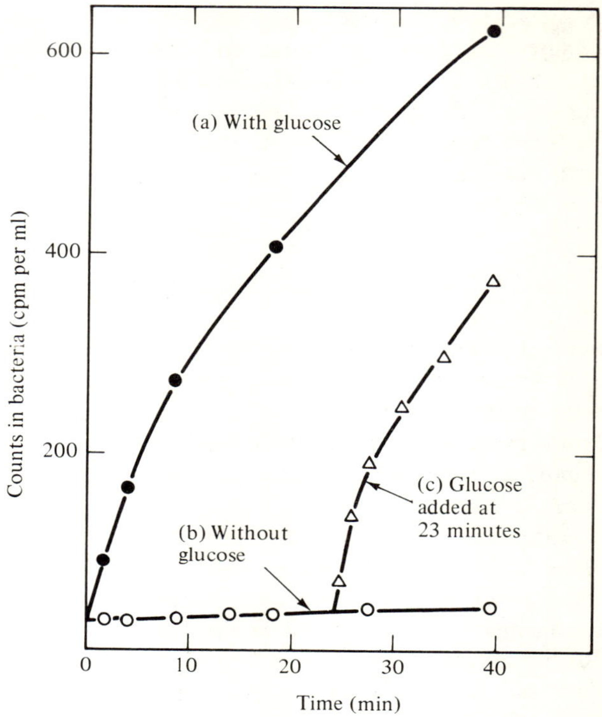

Fig. 54. Effect of glucose on uptake of ^{14}C-spermidine on *E. coli* strain B (412). The number of counts taken up in 40 min in flask A represents 0.0012 μmole per ml. This value may be compared with the value of 0.002 μmole of unlabeled spermidine originally present in the bacteria.

nisms for metallic divalent cations, e.g., Mg^{++}, has also been described (366).

Acetylation of Polyamines

When small amounts of ^{14}C-putrescine are offered to *E. coli* in exponential growth, over 95% of the ^{14}C in the cells appears in putrescine, spermidine and its acetyl derivatives (415). There are, therefore, no major paths of additional transformation of these compounds in the growing cells. Both forms of monoacetyl spermidine, as well as monoacetyl putrescine, are produced and excreted in chilled cultures (225). Diacetyl derivatives of these polyamines have not been detected, although spermine is converted to a diacetyl form (92). A similar conversion of putrescine to monoacetyl putrescine has been observed in staphylococci (330).

Increase of the pH of the medium from 6.8 to 8 increases acetylation of spermidine in *E. coli* strain B (92). When spermidine is present in the medium, the cells contain more spermidine, although the putrescine level is sharply reduced. As we might expect, far more acetylspermidine is elaborated. Table 10 shows that the presence of spermine in the medium

Table 10. Effect of Spermine on Amine Content of *E. coli* B (92)

		Amines				
Culture	*Spermine in medium* *	*Putrescine*	*Spermidine*	*Monoacetyl-spermine*	*Diacetyl-spermine*	*Spermine*
				μmoles/g wet weight		
1A	0	13	1.8	<0.01	<0.2	<0.02
1B	5×10^{-4} M	0.8	0.05	3.6	15	4.6
2A	10^{-3} M	1.0	<0.2	3.8	12.5	5.9
2B	10^{-3} M	0.2	<0.05	5.9	22	9.3

Cultures 1A and 1B were grown in parallel from the same inoculum from titers of 5×10^7 to 5×10^9 cells per ml. Cultures 2A and 2B represent serial samples from the same culture, grown from an initial titer of 10^8 cells per ml. Sample 2A was withdrawn at a titer of 2×10^8 cells per ml and 2B at a titer of 2×10^9 cells per ml. The levels of monoacetylputrescine and monoacetylspermidine A and B in cells grown in spermine were in all cases depressed to less than 0.2 μequivalent of amino groups per gram. Generation times were: 1A, 38 minutes; 1B 45 minutes; 2, 62 minutes.

* Initial concentration.

markedly reduces the content of both putrescine and spermidine in the cells while leading to an extensive synthesis of acetylspermine.

Although *E. coli* strain 15 TAU excretes large quantities of putrescine in the medium, whether grown on arginine or not, very little putrescine appears in acetylputrescine. On the other hand, *E. coli* 15 TAU converts significant amounts of spermidine to monoacetylspermidine in the absence of growth, and indeed this conversion appears to be a mechanism of excretion of excess spermidine.

The problem of the actual state and distribution of the amines between cells and medium has been confused (or clarified) recently by a startling result, presented in Fig. 55 (407). It appears that chilling *E. coli* B to 4° for a two-hour period leads to a significant conversion of cell spermidine (60%) to the acetyl form, whereas little is converted during exponential growth at 37°. In the absence of sensitive analytical methods or appropriately labeled compounds, the need to centrifuge large volumes of cultures to permit separation into cells and medium introduced a previously unsuspected problem. Some workers have solved this problem by filtration

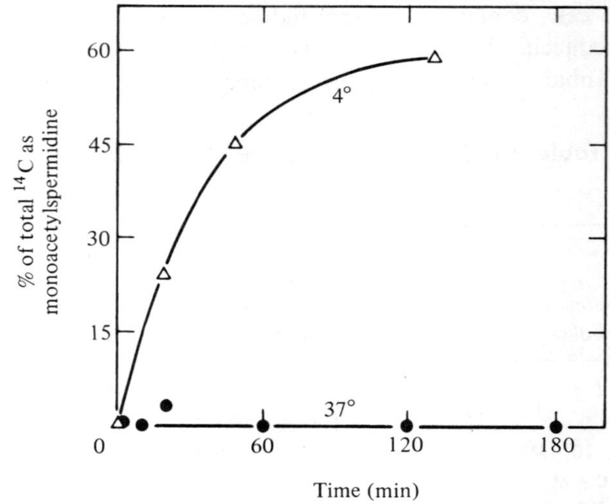

Fig. 55. Effect of temperature on the conversion of ^{14}C-spermidine to ^{14}C-monoacetylspermidine (407).

techniques. Our own approach has been to use the very sensitive dansyl procedure on acid extracts of small amounts of filtered cells or on small volumes of filtered media. For example, we can analyze 0.2 ml aliquots of media from cultures at 2–4 $\times$ 10^8 bacteria per ml. Analysis before and after acid hydrolysis reveals that a variety of organisms, such as relaxed strain TAU in the absence of uracil and phage-infected *E. coli* strain B, do indeed excrete acetylspermidine in situations other than exponential growth. Furthermore, even in exponential growth the presence of high concentrations of exogenous spermidine ($> 3 \times 10^{-6}$ M) results in a conversion of some cellular spermidine to the acetyl derivative (409).

It appears that under conditions of exponential growth, in the absence of exogenous polyamine or of precursor, cells do not normally synthesize an excess of polyamine and do not excrete it. Efficient control mechanisms appear to restrict such wasteful production. However, under the unusual circumstance in which the presence of an exogenous precursor can result in unusual accumulations, polyamine can be excreted. Thus, in exponential growth of *E. coli* 15 TAU in the presence of arginine, we find large amounts of putrescine in the medium even at 37°. As is seen later (Fig. 66), cells in which K^+ is largely replaced by Na^+ enormously increase putrescine synthesis, and most of this diamine is excreted into the medium at 37° (333). Excretion of acetylspermidine can also be observed in cells in which distortions occur in the pattern of nucleic acid synthesis. Thus, in a relaxed strain of TAU making large amounts of spermidine, an arrest of RNA synthesis will permit a continuing synthesis of spermidine, most of which is excreted as the monoacetyl derivative (85).

The mechanism of cold-activation of acetylation is obscure. Little is generally known about the control of acetylation of polyamines, although acetylation may prove to be significant in controlling polyamine function.

Polyamines and RNA Synthesis

I have been concerned with the problem of the nature of the linkages of the nucleic acids to proteins for about thirty years (56, 70, 61, 71, 65) and have continued to wonder about the nature of the cations that neutralize the nucleates in cells. My preoccupation was further stimulated by the discoveries of large amounts of the polyamines (156, 8, 7) in the phages with which I was working. The discovery that a bacterial extract could be fractionated by the cationic streptomycin into various cellular elements differing in the nature of their nucleic acid (67) led us to ask whether these cellular organelles contained some of the organic cations. Having observed the effect of ionic strength in reducing the association of streptomycin with nucleates, we prepared extracts at low ionic strength, a condition that leads to the dissociation of *E. coli* ribosomes into subunits. As noted earlier, some years ago Lichtenstein and I observed that a significant amount of cell spermidine is found in isolated ribosome subunits and that indeed exogenous spermidine can cause the reassociation of these subunits (66). It was shown that Mg^{++} and spermidine, both ribosomal components, acted together in producing this effect, and we suggested that they have this role within the cell.

We then began to ask if spermidine did not fulfill an important role in the biosynthesis of ribosomal RNA, which constitutes the largest part of cell RNA. This inquiry was started by our observation that the cationic antibiotic, streptomycin, stimulates synthesis of ribosomal RNA, concomitantly with the development of lethality (395). In the course of testing the correlation between the lethality caused by streptomycin and the stimulation of RNA synthesis, which are rigorously tied together in all the systems we have studied, we developed a number of bacterial systems in which RNA synthesis could be controllably varied (305, 309, 65). This correlation has been exploited recently by Russian workers in the selection by streptomycin of mutants containing a temperature-sensitive RNA polymerase (203a).

Using *E. coli* strain 15 TAU, we demonstrated that the increase of RNA and intracellular spermidine showed similar patterns in exponential growth in a complete medium or in the presence of chloramphenicol, streptomycin, or both antibiotics. Such results are presented in Fig. 56. On the other hand, putrescine was selectively lost from the cell as a result of streptomycin action (305, 309). Thus, the distributions of the two polyamines were dissimilar, spermidine being associated with cell structure

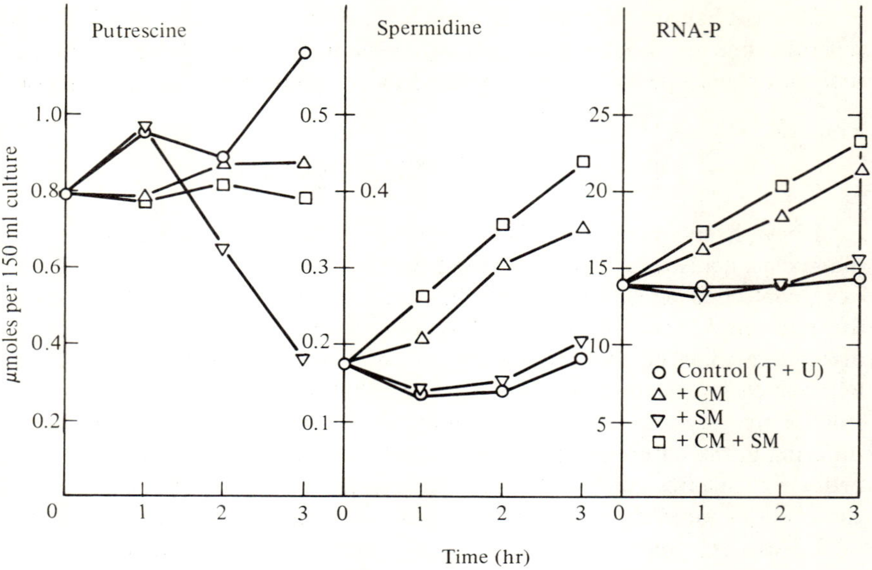

Fig. 56. Effect of chloramphenicol and streptomycin on the intracellular accumulation of unconjugated polyamines and RNA in *E. coli* 15 TAU during amino acid starvation (309). T = thymine; U = uracil; CM = chloramphenicol; SM = streptomycin.

under conditions in which putrescine existed in a rapidly dissociable soluble pool.

Similar results were obtained in a comparable experiment done in the absence of the required amino acid, arginine (305, 309). We have shown that this condition produced a decrease of 85–95% in uracil uptake, a condition defined as the "stringent" amino acid control of RNA synthesis (396). Despite the absence of protein synthesis in this system, streptomycin was nevertheless lethal and produced a slow but significant concomitant stimulation of uracil incorporation (395).

The presence of chloramphenicol also stimulates RNA synthesis and the accumulation of cellular spermidine (309). As remarked earlier, in the absence of the amino acid, spermidine continues to be made but is found entirely in the medium as acetylspermidine. When strain TAU is incubated in the absence of thymine, RNA is synthesized and spermidine accumulates in the cells at the same time, although DNA is not synthesized.

Effects of Methionine on RNA Synthesis

It has been reported by Ron and Davis (327) and is shown in Fig. 57 that the addition of methionine to growing cultures of many strains of

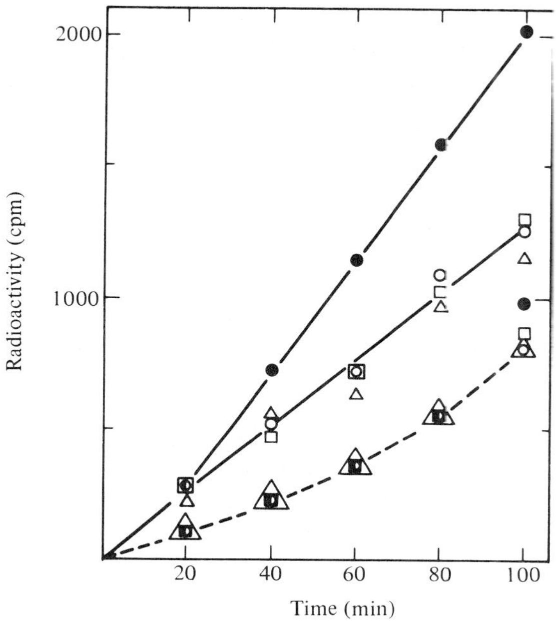

Fig. 57. Effect of methionine, histidine, and threonine on RNA and protein synthesis in *E. coli* 15 T⁻A⁻ (327). ——— RNA synthesis; - - - - protein synthesis; -O-O- no addition; -●-●- methionine; -□-□- threonine; -△-△- histidine.

E. coli stimulates synthesis of ribosomal RNA preferentially In confirmation of this fact, we have shown that in strain TAU, exogenous methionine not needed for growth produces a 20% increment in RNA. Methionine also permits a 75% increase in total cell spermidine, with scarcely any change in putrescine (309).

In our studies with the relaxed K12 strain requiring both methionine and leucine for growth, we have observed that the presence of methionine alone permits a greater synthesis of RNA than with only leucine (309). As we would expect, the presence of methionine alone allows a marked synthesis of spermidine. I shall shortly comment on the problem of the biosynthesis of RNA in the absence of an essential amino acid.

A more direct demonstration of methionine as a precursor of spermidine active in determining cellular function has recently been described by Witkin and Rosenberg (459). The bacterium, *Myxococcus xanthus*, commonly found on decaying material, has been observed to undergo a special type of morphogenesis depending on the availability of methionine or polyamine (459). Under nutritionally limiting conditions, such as methionine starvation, cells aggregate to form a fruiting body within which the cells become refractile microcysts. Exogenous putrescine increased microcyst formation. Exogenous spermidine (2 mM) prevented microcyst formation induced by methionine starvation with or without putrescine.

Sublethal concentrations of azide or cyanide in strain 15T⁻ produce a preferential inhibition of RNA synthesis, and this inhibition is specifically relieved by methionine (280). The obvious question of the relations of the polyamines to this set of effects has not been explored.

Effects of Spermidine on RNA Synthesis

Having observed the simultaneous accumulation of RNA and spermidine, we asked whether the two functions are independent, whether RNA synthesis controls spermidine synthesis, or whether spermidine helps to control RNA synthesis. As a first approach, we added exogenous spermidine to the stringent strain of 15 TAU in the absence of arginine (305, 309). At high concentrations, required we think to overcome the permeability barrier described earlier, we observed a stimulation of RNA synthesis, as in Fig. 58 (309). The newly synthesized RNA is found in a peak ascribed to

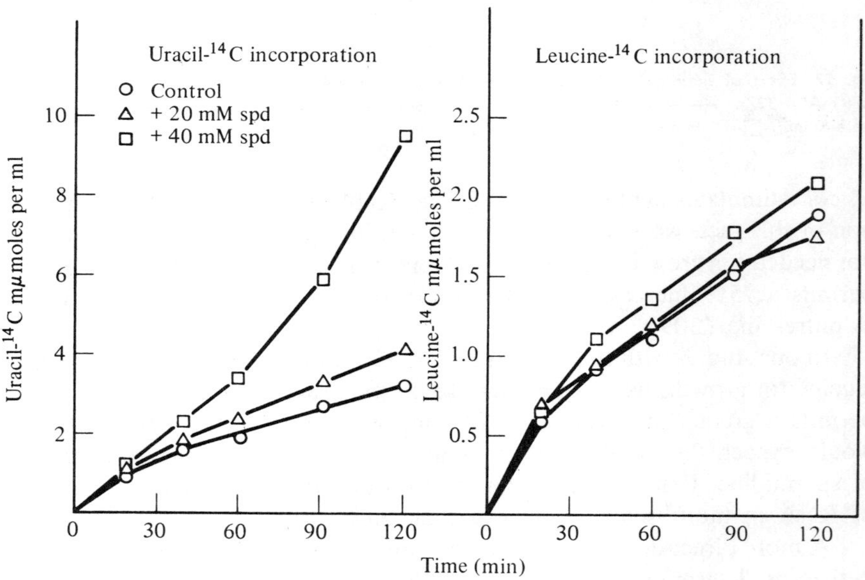

Fig. 58. Effect of exogenous spermidine on the incorporation of ^{14}C-uracil and ^{14}C-leucine into acid-insoluble material in *E. coli* 15 TAU during arginine starvation (309).

incomplete ribosomes. The bulk of such RNA detected after deproteinization of the extracts is in types of RNA characteristic of the ribosomal units and does not appear to contain transfer RNA.

The stimulatory concentration (40 mM) presented in Fig. 58 is not lethal, nor does it alter the low rate of incorporation of exogenous ^{35}S or of leucine-^{14}C in the absence of arginine. I stress this fact because of the report of Ezekiel and Brockman (104), who from data very similar to those of Fig. 58 [see Figs. 4 and 5 (104)] concluded that the relaxation of RNA synthesis was related to an inhibition of protein synthesis and an increased turnover of amino acids. The unchanged low rate of leucine-^{14}C incorporation under conditions of stimulation of RNA synthesis does not support this interpretation. Sussman and Gilvarg have recently reported that, in fact, relaxed strains have considerably less protein turnover than do stringent strains (402). Although the presence of an inhibitor of protein synthesis such as chloramphenicol frequently stimulates RNA synthesis, the mechanism of this effect is quite obscure in general. We shall see soon that there is no evidence to prove that the action of chloramphenicol is identical to that of stimulatory concentrations of spermidine.

In these experiments it was shown too that exogenous putrescine prevented the stimulation of synthesis of RNA produced by exogenous spermidine (305). It was suggested that the ratio of these potentially competing substances, rather than the concentration of spermidine alone, determines the stimulatory response. That such a balance appears to operate in bacterial division (179) and microcyst formation (459) has been noted earlier.

We have sought to find systems in which the effects of spermidine on RNA synthesis can be found at more physiological concentrations, as in the effect on spermidine in preventing microcyst formation. In one *E. coli* system, first described by Simon (372), the morphine analog, levorphanol, was shown to inhibit synthesis of ribosomal RNA. An exogenous concentration of 1.5 mM spermidine but not of putrescine prevented this effect almost completely. If we take an organism inhibited by levorphanol and add spermidine, as in Fig. 59, we can see that RNA synthesis resumes immediately (371). Levorphanol produces a rapid change in membrane permeability which under conditions of continuing metabolism results in a rapid loss of putrescine and some other ions from *E. coli* (371, 371a).

It may be mentioned that levorphanol has recently been reported to inhibit the synthesis of ribosomal RNA in the nucleoli of rat liver (342). No effects of levorphanol were detected on synthesis of messenger RNA or of transfer RNA.

On Relaxed Organisms

The stringent or relaxed control of RNA synthesis was shown by Stent and Brenner (394) to map as a single genetic locus, the RC locus. If the

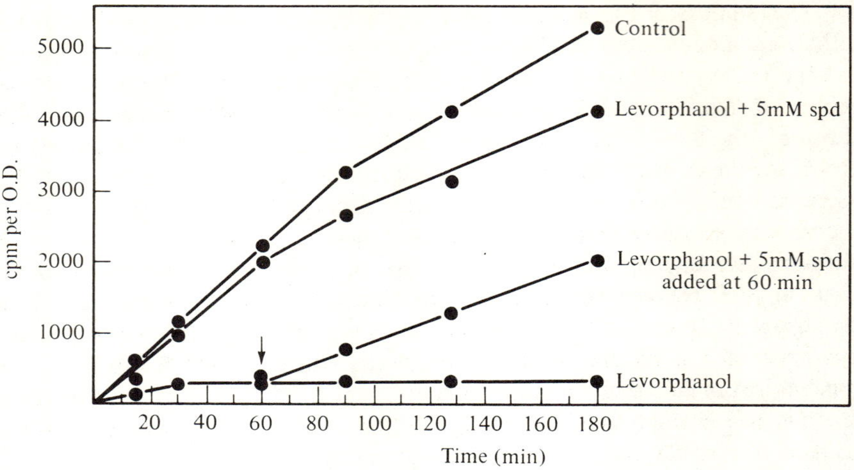

Fig. 59. Reversal of levorphanol inhibition of RNA synthesis by spermidine in *E. coli* K13 (371).

phenomenon of relaxed control of RNA synthesis is related to spermidine concentration, changes in the polyamines should be demonstrable in organisms relaxed in RNA synthesis in the absence of a required amino acid, i.e., as a result of the presence of the RC^{rel} allele. Moreover, we should think that a methionine-deficient organism would not be relaxed. Our analyses of relaxed organisms, such as the relaxed K12 strains, have revealed that in contrast to other strains, these cells have a fairly high content of spermidine and, in fact, may contain more spermidine than putrescine. It can be imagined that an intracellular redistribution and turnover of this spermidine could account for the net synthesis of RNA. Undoubtedly, these questions warrant additional study, but they appear intrinsically more difficult than do other approaches.

We have bypassed this question by studying an isogenic pair of stringent and relaxed strains of *E. coli* 15 TAU. Fiil and Friesen (109) in Copenhagen derived the relaxed mutant from the stringent parent that we had first isolated in 1955. The selection procedure permits relaxed mutants in the stringent culture to make RNA in the absence of the amino acid. The accumulation of RNA appears to inhibit subsequent growth of the mutants in a complete medium, a phenomenon first described by Borek (37). The stringent cells are selectively killed by penicillin and surviving relaxed mutants are thereby concentrated. Thus, the survival and selection of relaxed strains may be determined by this phenotypic expression itself, i.e., RNA synthesis.

The isogenic pair of stringent and relaxed strains of TAU were shown to be identical in their amino acid deficiency; the RNA synthesis in the relaxed cells, therefore, is not due to an increase in arginine supply or in protein turnover (65). Both strains suffer thymineless death at the same rate; however, the relaxed strain is killed much more rapidly in streptomycin, a phenomenon that appears once more to relate to the phenomenon of stimulated RNA synthesis (65). Even the killing of the stringent strain by the antibiotic in the absence of arginine can be hastened by a supply of methionine, as can the stimulation of RNA synthesis in such conditions (65).

We have compared the polyamine contents of the cells in growing cultures and have shown that at comparable turbidities the relaxed cell has more RNA (+ 17%), spermidine (+ 39%) and putrescine (+ 39%) than the stringent cell (65). Of course, these intracellular differences become markedly exaggerated in the absence of arginine when uracil is supplied to permit RNA synthesis, as can be seen in Fig. 60. Of great interest

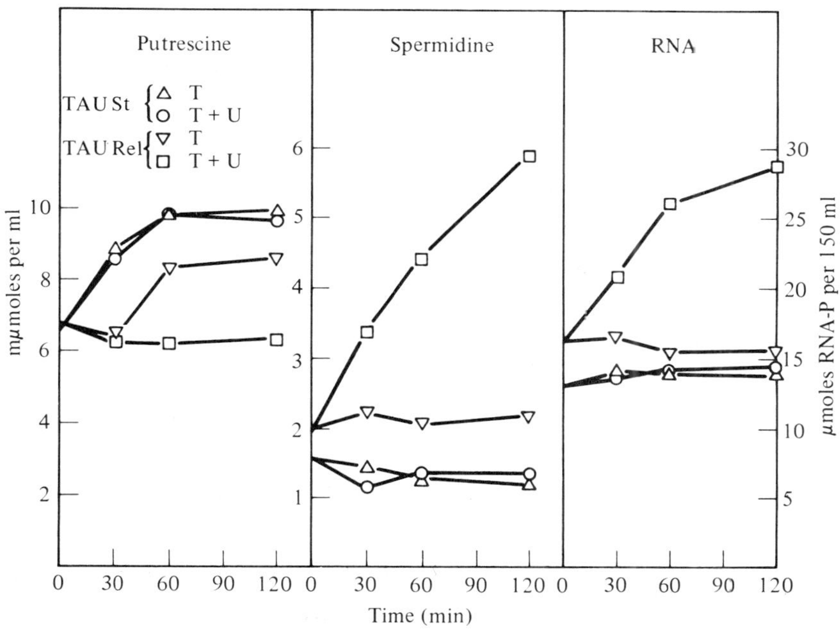

Fig. 60. Accumulation of RNA and free polyamines within cells of TAU Rel and TAU st (65). T = thymine; U = uracil.

is the absolute shut-off of accumulation of intracellular putrescine in these conditions. No spermidine builds up in the cell in the absence of RNA accumulation. This condition is achieved in relaxed cultures by omitting exogenous uracil (65).

So far, the analysis merely extends our data in familiar directions. However, let us examine total spermidine synthesis, i.e., spermidine in medium plus cells, in the relaxed culture lacking amino acid, as in Fig. 61. In the presence or absence of RNA synthesis, we see that the same total amount of spermidine is made in both conditions. Furthermore, the rate of synthesis in the relaxed cell is about three times that of the stringent cell. In addition, the synthesis of putrescine in the relaxed cell is inhibited when

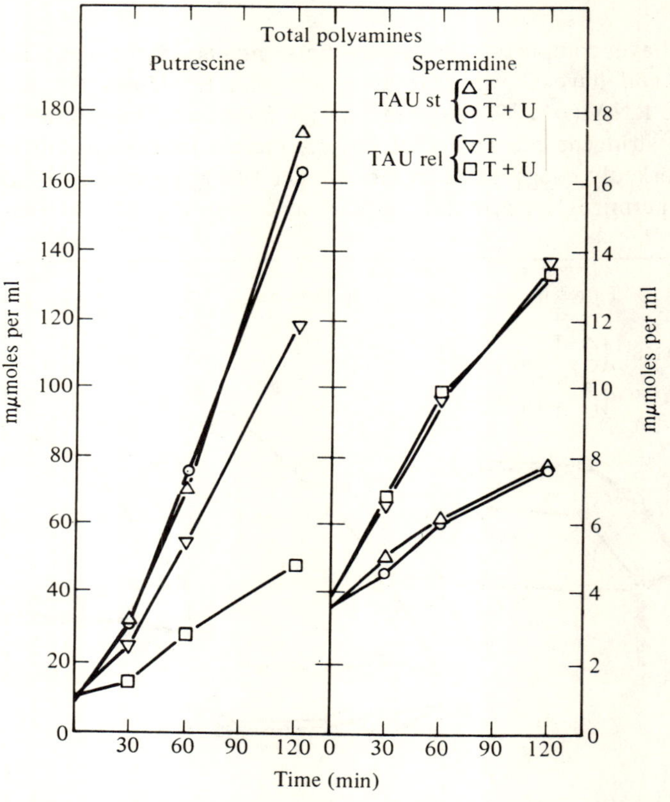

Fig. 61. The total production of putrescine and spermidine by TAU rel and TAU st incubated in the absence of arginine (65). The experiment is that presented in Fig. 60. Aliquots were incubated in the presence or absence of uracil, controlling RNA accumulation within the cells. At zero time, all cultures contained 2.7–2.8 $\times$ 10^8 viable cells per ml.

RNA and spermidine accumulate; this synthesis is about one-third that of the stringent cultures. Thus, in the relaxed cell, which was presumably selected on the basis of its accumulation of RNA, the ratio of synthesis of spermidine to total polyamine under conditions of relaxed synthesis was about nine times that in stringent cells. Since an exaggerated synthesis of spermidine occurs whether RNA is made or not, we suggest that the abnormality in polyamine biosynthesis and utilization is more directly related to the mutation in the RC gene than is RNA synthesis itself. To put it another way, we suggest that the aberration in the control of RNA synthesis may be causally related to the prior abnormality in polyamine synthesis. The enzymatic nature of the latter abnormality has not yet been clarified.

On the Nature of Relaxed Control

It is apparent that our study of RC^{rel} bacteria has not revealed any facts inconsistent with the mass of existing data implicating the availability of spermidine as a controlling element in RNA synthesis. We know that relaxed bacteria have several unusual properties, of which altered metabolic capabilities (390) and increased permeability (253) are two. A recent result contradicting earlier reports and indicating that lipid synthesis is not under control of the RC gene has been interpreted to suggest that the control of RNA synthesis cannot be simply at the level of availability of nucleoside triphosphates (437). The nature of the primary effect of the RC gene has not yet been demonstrated, although it appears quite possible that the aberration in polyamine biosynthesis could result from the change in the RC locus. There appear to be no differences in the RNA polymerase in the two types of organism (287), nor is the inhibition of RNA synthesis by nonacylated tRNA considered to be determining (270), at least in the form of the hypothesis initially proposed by several groups of workers (394, 215, 434). Alternative hypotheses continue to be investigated, nevertheless, and many observations (difficult to interpret) have been reported (96). Among one of the interesting recent findings is the discovery of the elaboration of an unusual guanine nucleotide, ppGpp, produced uniquely under stringent conditions (53), but the significance of this observation is not yet clear.

I do not intend to discuss all of the current hypotheses of stringency and relaxation. However, if I suggest a hypothesis of relaxed control based on polyamine synthesis, it seems appropriate to present one or more mechanisms whereby these two phenomena could be causally related.

The accumulation of rRNA observed in relaxed biosynthesis is a consequence of both synthesis and degradation. As shown by Nierlich (281), the uptake of a base from the medium will not exceed the net rate of entrance of nucleotides into stable RNA. Various workers have shown that changes of pool nucleotides cannot account for the balance obtained. Additional data of Nierlich have suggested that a stringent cell makes RNA at only a slightly lower rate than does a relaxed cell. It appears that the degradation of the RNA in stringent cells may be occurring at a rate that prevents net accumulation (282). Nierlich's data are entirely consistent with the possibility that a relaxed cell may be degrading RNA less rapidly than a stringent cell. A decade earlier, Herbst and Doctor had shown that *Hemophilus parainfluenzae,* which requires a polyamine for growth, rapidly loses RNA into the medium as acid-soluble materials when the organism is incubated in growth medium without polyamine (151). *H. parainfluenzae,* initially grown on the polyamine, putrescine, will not lose its RNA if spermidine or spermine are added to the medium.

The RNA accumulated in stringent *E. coli* treated with chloramphenicol is degraded when chloramphenicol is removed. It has been shown by Lazzarini and Santangelo (220) that both synthesis and degradation occur in the presence of chloramphenicol, and both processes are still going on at the time that net accumulation stops. Mills and Dubin (261) have demonstrated that addition of spermine to such a system when accumulation is almost stopped permits an immediate stimulation of accumulation, as shown in Fig. 62. Thus, the mechanisms of stimulation of RNA synthesis by

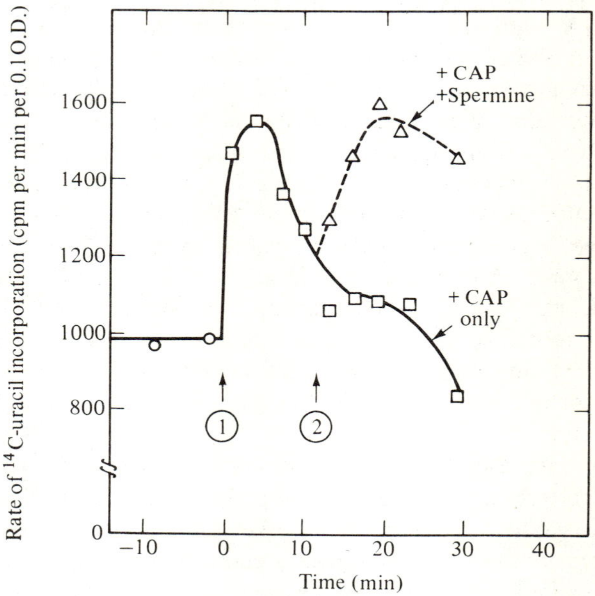

Fig. 62. Stimulation of RNA synthesis by spermine in chloramphenicol-treated cells (261). At arrow (1), chloramphenicol was added to a growing culture. At 12 min, an aliquot was added to a tube containing spermine.

chloramphenicol and by exogenous polyamines appear to be different, since the effect of polyamine in stimulating accumulation can be demonstrated after a maximal chloramphenicol effect (500 μg per ml) on RNA synthesis has been expressed.

A similar effect is demonstrable in relaxed TAU in which turnover of RNA is still active when net accumulation, measured as incorporation, stops. The addition of exogenous spermidine permits the accumulation of larger amounts of RNA in cells that had never been making protein. The simplest interpretation of these data is that the organic cations stabilize newly formed RNA, i.e., reduce degradation, and permit the establishment of new levels of accumulation before the rates of synthesis and degradation become equivalent. Data showing that the RNA molecules of relaxed cells contain more spermidine than do those of stringent cells are described in Chapter 4. The apparent stimulation of RNA synthesis by streptomycin may derive in significant part from the stabilization of newly synthesized RNA, although it has never been demonstrated. The nature of the stimulation by chloramphenicol remains obscure.

It may be mentioned that the stability of intracellular RNA accumulated during chloramphenicol treatment of a stringent culture will vary as a function of incubation conditions following removal of the antibiotic. Putrescine at high concentrations does not prevent and may stimulate such degradation; spermidine prevents much of this enzymatic action (38). In the presence of proflavine, which inhibits transcription, putrescine does not permit RNA synthesis, whereas spermidine does so, suggesting an additional action of high exogenous concentrations of spermidine at the level of transcription.

Our recent findings that putrescine or spermidine permits DNA synthesis in T4 infection of a putrescine-deficient mutant of *E. coli* inevitably raises the question of whether the polyamine facilitates DNA replication or stabilizes newly made DNA. This question also arises secondarily as a result of the recent work of Donini (89), who has shown that DNA synthesis in a T4-infected stringent cell is under amino acid control but escapes such control in an infected relaxed cell. The first question might be: does the escape of DNA synthesis from amino acid control in relaxed cells reflect a function of the unusual production of spermidine in such synthesis, an overproduction known to occur in some relaxed cells?

Polyamines and Extension of RNA Chains

Winslow and Lazzarini (458) think that Nierlich has overestimated RNA synthesis in stringent organisms by a factor of two to three. They believe that there is a greatly reduced rate of chain elongation during stringency. If these workers are in fact describing the major aspect of stringent control, it becomes necessary to consider the role of polyamines in this context.

That the polyamines also stimulate chain elongation in systems containing DNA-dependent RNA polymerase in vitro is well known. In 1963 Krakow (209) described the stimulation of synthesis by polyamines when double-stranded DNA was used; this effect was not observed with denatured DNA (211) or with synthetic polyribonucleotides. The same result with the RNA polymerase of *Azotobacter vinelandii* was obtained by Fox and Weiss for the enzyme from *Micrococcus lysodeikticus* (114). These workers showed that with the latter enzyme spermidine does not stimulate RNA-primed reactions (113). These observations are sufficiently widespread to have warranted incorporation of spermidine into many recipes of assay for many types of DNA-dependent RNA polymerase, i.e., bacterial, plant, and animal polymerase.

A careful analysis of the action of *E. coli* RNA polymerase on double-stranded templates by Fuchs et al. (122) has revealed that the extension of RNA chains requires a high ionic strength; synthesis stops at low ionic strength, less than 0.2, due to a first-order inactivation of the synthesizing enzyme (Fig. 63). This arrest can be prevented if spermidine and Mg^{++} are

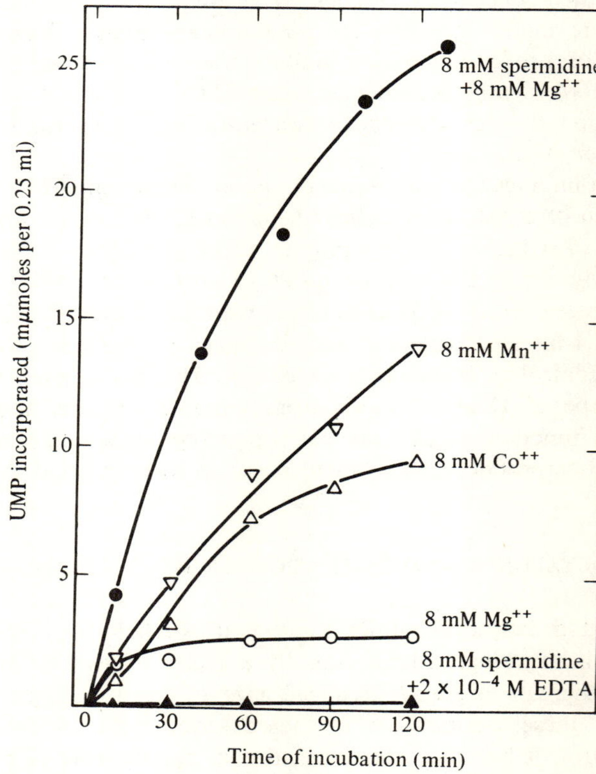

Fig. 63. Effects of cations on the kinetics of synthesis by *E. coli* DNA-dependent RNA polymerase (122).

both present. In other words, spermidine cannot entirely replace the Mg^{++} requirement. The arrest of synthesis can also be reversed immediately by addition of Mg^{++} or spermidine (note Fig. 64).

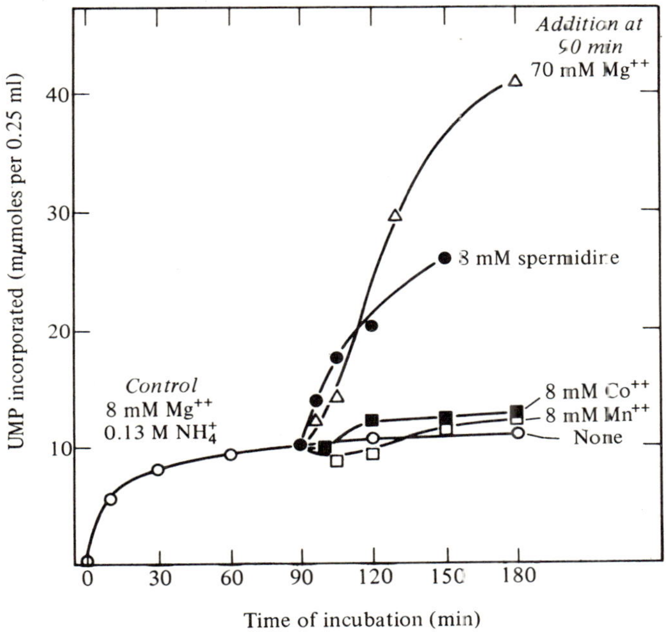

Fig. 64. Effect of adding Mn^{++}, Co^{++}, or spermidine in the plateau of RNA synthesis (122).

Subsequent work of Millette (260) using α-^{32}P-ATP to study elongation and β,γ-^{32}P-ATP as a measure of chain initiation has revealed that both functions stop together at low ionic strength. At high ionic strength, initiation and chain extension continue (388, 222, 245) with RNA being released from the reaction complex much more rapidly (260, 245).

The experiments of Petersen et al. (296), using appropriately labeled ATP, have also shown that spermidine does stimulate chain initiation as well as chain extension on a native double-stranded DNA template. On the other hand, the polyamine has little effect on these activities on a denatured DNA template. This result has supported the suggestion that the polyamine increases available initiation sites on native DNA by repairing the secondary structure of defective single-stranded regions of DNA. This as well as other possibilities have been summarized by Abraham (1) and by Stevens (399). However, among these hypotheses, the soundest appears to be that polyamines facilitate the elimination of inhibitory RNA from the DNA-enzyme-substrate complex.

The studies of Krakow (209, 211) and of the Chicago group (114,

113, 112) have shown that the presence of RNA chains on the complex do in fact inhibit further synthesis. As described by Krakow (210), the addition of ribonuclease permits continuation of the reaction. Maitra and Hurwitz (246) have also demonstrated increased initiation by RNase in this system. We have seen that a similar effect is demonstrable with 1–2 mM spermidine, a result that goes hand in hand with improved release of RNA from the complex (112). However, it is not clear if the formation of spermidine salts of the RNA, while the RNA is still attached to the enzyme-template complex, does not also minimize the inhibitory action of the nascent RNA chains. Also, the possible relations of spermidine and the psi termination factor in the polymerase reaction are still to be clarified.

The demonstrations that the polyamines have these effects in vivo and in vitro at approximately physiological concentrations are no more than demonstrations that the stimulation of transcription and the reduction of degradation are consistent with the phenomena of relaxed control. The effects of these compounds cannot be discarded out of hand; their elimination from consideration will require at least as serious an effort as that which implicates them in the phenomenon. On the other hand, the complexity of the problem does not yet permit a simple demonstration of a 1:1 causal relation.

We may close this section by recording a third effect of spermidine on RNA polymerase; this is the effect of spermidine in strand selection during transcription. Gumport and Weiss (142) have indicated that with certain DNA templates, such as the circular double-stranded replicative DNA of ϕX174, the RNA polymerase of *M. lysodeikticus* or of *E. coli* produce RNA complementary to both strands. Such RNA will hybridize with itself. The addition of spermidine to the transcription mixture sharply increases the asymmetry of transcription, as presented in Table 11. This result is due

Table 11. Effect of Reaction Conditions on the Symmetry of Transcription (142)

Reaction conditions	*cRNA Synthesized (μg)*	*RNA-RNA hybrid (%)*	*DNA on filter*	*DNA-RNA hybrid (%)*
I. $MgCl_2$	3.55	62	RF	50
			SS	12
II. $MnCl_2$ + spermidine	4.10	35	RF	48
			SS	7
III. High $MgCl_2$ + spermidine	4.82	18	RF	67
			SS	3
IV. EC RNA polymerase, $MnCl_2$, + spermidine	14.10	14	RF	66
			SS	2

to an effect on the enzyme itself, rather than a repair of denatured regions in the template.

The transcription of T4 DNA by these bacterial enzymes was essentially asymmetric under various conditions of ion supply. However, the cation content of such DNA derived from a polyamine-containing phage is never defined in experiments with T4 DNA. It is usually stated that such DNA has been purified by phenol extraction, although it is known that phenol extraction at low ionic strength does not remove polyamine from DNA (8, 7).

Possible Interrelations of Polyamines and Carbohydrate Metabolism

Sanwal has recently reported (345) that spermidine can replace Mg^{++} as an activator of *E. coli* glucose-6-phosphate dehydrogenase (351). In this reaction, spermidine has an apparent K_m of 0.2 mM, in contrast to an apparent K_m of 2 mM for Mg^{++}. Since 0.2 mM is well below the apparent concentration of spermidine in *E. coli,* this polyamine may well be the natural activator for this key step. Spermidine does not replace Mg^{++} in the activation of 6-phosphogluconate dehydrogenase.

Many years ago we showed (219, 232) that the ribose of RNA in *E. coli* is derived mainly from the phosphogluconate path of which this enzyme catalyzes the first step. It therefore appears possible that a compound facilitating RNA synthesis at the level of transcription is also stimulating the supply of RNA precursors, as suggested in the schema in Fig. 65. Obviously, such a hypothesis warrants our serious examination.

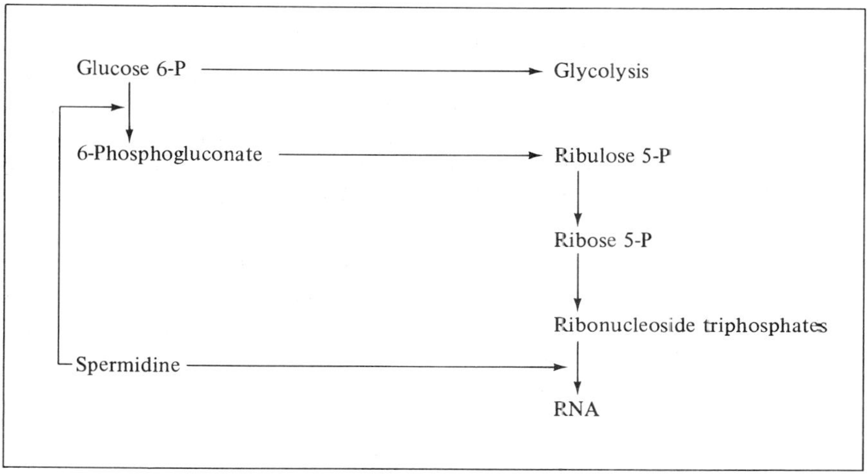

Fig. 65. Postulated sites of activation by spermidine in promoting synthesis of RNA.

There are several possible systems for exploration of this possibility; putrescine-deficient mutants and the stringent and relaxed isogenic pairs may be two. Studies of the latter system have revealed small changes in the balance of paths of glucose metabolism under conditions in which spermidine is made in the relaxed cell (54). These studies are continuing.

Another system we have been studying in some detail involves a strain of *E. coli* unable to accumulate potassium ion (100). The organism requires high levels of K^+ for growth. When transferred to Na^+, the cells are unable to synthesize protein but do make RNA (233). We have found that in the Na^+ medium, the cells make unusually large amounts of putrescine (333) and appear to provide another potentially useful system for dissecting interrelations of the polyamines and RNA synthesis.

ARE AMINES HELPING TO REGULATE CELLULAR IONIC BALANCE AND pH?

The K^+-deficient *E. coli* strain used in studies mentioned above is essentially retention-negative (234), unlike some others deficient in uptake of K^+ or possessing other types of K^+ dependency (101). K^+ not only plays an important part in osmoregulation in these bacteria (102) but also activates many cell enzymes (401). The strain is stringent in media containing K^+ and lacking an amino acid, but the synthesis of RNA is relaxed in Na^+ with or without amino acids (54). This RNA synthesis in Na^+ may be increased several-fold (233) by the appropriate addition of adenosine and guanine (333). K^+ has been shown to be essential to the formation of purine and in the conversion of inosinate to xanthylate (299). Na^+ inhibits the insertion of adenine to form adenylate (30). K^+ is also reported to be essential in yeast (272) for the activation of methionine to form S-adenosylmethionine, essential for the synthesis of spermidine. In addition, the cation activates a carbamylphosphate synthetase (249); in apparent confirmation, we have observed that exogenous uracil also stimulates RNA synthesis in this system (333). Thus, maximal RNA synthesis would be expected and is indeed obtained from exogenous addition of adenosine, guanine, and uracil (333).

We have asked if spermidine synthesis is in fact eliminated in the Na^+ medium. A significant net accumulation is not observed; however, a study with 2-^{14}C-methionine has revealed a small formation of S-adenosylmethionine, as well as a small synthesis of spermidine, perhaps 25% of normal. In a possible explanation of such an effect, Lubin (233) observed that in Na^+-washed cells about 5% of their normal K^+ content (about 10 mM) cannot be washed out. In any case, the synthesis of RNA via RNA polymerase in this system does not require the high level of K^+

which is normally present, nor does it require a high level of newly synthesized spermidine.

Beset by this apparent contradiction of our generalization on the relation of polyamines to RNA, we discovered that in this organism in Na^+ medium, putrescine biosynthesis had now increased enormously, about 8–10-fold (333). Studies of urea production in the presence or absence of arginine revealed that the agmatine route to putrescine is minor in this organism; putrescine stems almost entirely from ornithine and ornithine decarboxylase in all conditions of biosynthesis. Since no protein is being made, the system had not been derepressed. However, exogenous addition of spermidine or even of methionine markedly reduces the production of putrescine (as shown in Fig. 66); therefore, a relative lack of spermidine

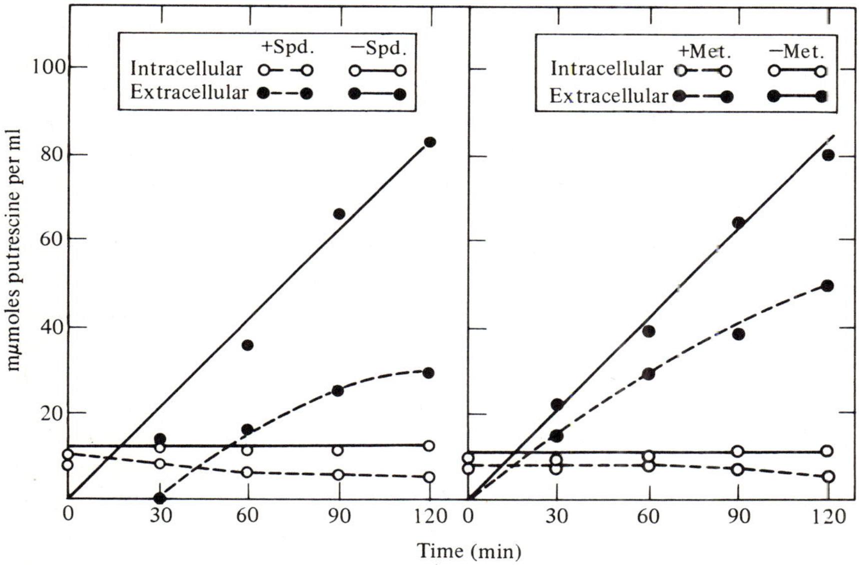

Fig. 66. The synthesis and excretion of putrescine in *E. coli* B 697 suspended in a Na^+ medium in the presence or absence of spermidine or methionine (333).

appears to have eliminated a controlling feedback inhibition. It is not yet known whether this effect will account for the entire stimulation of putrescine biosynthesis or whether additional contributing factors include an inhibition of normal enzymes by K^+. An increase in the supply of ornithine does not account for the difference between the effects of K^+ and Na^+ on putrescine synthesis.

Although spermidine production in this system is severely depressed, putrescine synthesis has been enormously expanded, and we must now ask if putrescine at a sufficiently high local concentration cannot fulfill the role of both K^+ and spermidine in facilitating the maintenance of transcription. We have found that tRNA made in K^+ contains spermidine and is low in putrescine. However, the transfer RNA fraction isolated from bacteria incubated in Na^+ was demonstrated to contain both species made during this incubation and larger amounts of putrescine than does normal tRNA. The physiological significance of such an effect has not been explored, but it suggests that, when K^+ is deficient, putrescine may assume some of the roles of spermidine in nucleic acid structure and function.

In studies of potassium metabolism in *Aerobacter aerogenes* growing in a chemostat (433), K^+, Mg^{++}, and phosphorus contents were stoichiometrically related at all times, although the bacterial contents of each were functions of the growth rate. Since RNA is the only known macromolecule whose concentration is a function of growth rate, we surmise that the three elements are fulfilling a defined role in RNA structure. We definitely suspect that this is also the case for spermidine. In these terms, the increase of putrescine in cells deficient in both K^+ and spermidine suggests that newly formed RNA and putrescine may now be associated. We think that we have detected this result in newly synthesized tRNA in this system.

Marked abnormalities of polyamine metabolism in plants have been known for many years. In 1952 Richards and Coleman (318) showed that putrescine accumulates in potassium-deficient barley and other plants and that the excess of polyamine seems to induce a necrosis in various species (73). Smith (381) has recently demonstrated the presence of putrescine, spermidine, and spermine in the leaves of many higher plants. He has also estimated the very high levels of putrescine and polyamines produced in the leaves of six species grown in potassium-deficient conditions. Agmatine accumulated in various plants too, and this substance was shown to be derived from arginine, which itself appeared to be the major precursor of putrescine in the plants (382). However, the route of degradation of agmatine in many of these organisms involves the formation of carbamylputrescine, as presented in Fig. 12. Carbamylputrescine also may be formed by the decarboxylation of citrulline, a reaction that has been detected in extracts of K^+-deficient plants but not in extracts of normal plants (75).

In an important series of studies, Smith has described the plant enzymes, L-arginine carboxylase (or decarboxylase) (378), an agmatine iminohydrolase (379), and an N-carbamylputrescine amidohydrolase (380). He has studied the relation of the levels of these enzymes to the growth of the plants in potassium-deficient soil. The decarboxylase at least doubles its activity in deficient plants (378), whereas the iminohydrolase is not significantly increased under conditions in which agmatine had increased 8-fold and putrescine 30-fold (379). Finally, the activity of the amidohydrolase was considerably higher in deficient leaves. Although the result of deficiency is similar in plants and bacteria, the exaggerated synthesis of putrescine is stimulated by K^+ deficiency in plants along a path that is quite different from that in bacteria and involves responses of two enzymes not implicated in the bacterial route. It has been found that putrescine production is also greater in tobacco grown in sulfur-deficient soil (374, 463), thereby suggesting a possible effect via the resulting methionine and spermidine deficiencies. I should mention that the pyrrolidine ring of the nicotine ring has been shown to have arginine, agmatine, carbamylputrescine, and putrescine as precursors (464).

According to the study by Sinclair (374), increases of agmatine in plants were observed in a variety of mineral deficiencies, whereas K^+ deficiency related more specifically to putrescine accumulation. It was suggested as a result of this study that the accumulation of amines represented a homeostatic mechanism for maintenance of ionic balance. We have seen an apparent prototype of this type of effect in the substitution of spermidine or hydroxyputrescine for Mg^{++} deficiency in appropriate bacteria.

Smith and Sinclair (383) have also proposed that amine formation and excretion may serve as an internal compensating mechanism to keep intracellular pH fairly constant. They showed that barley seedlings responded to the feedings of acid to the roots by increasing both arginine decarboxylase and N-carbamylputrescine amidohydrolase in the leaves, as presented in Fig. 67. The arginine, agmatine, and putrescine contents of the leaves were

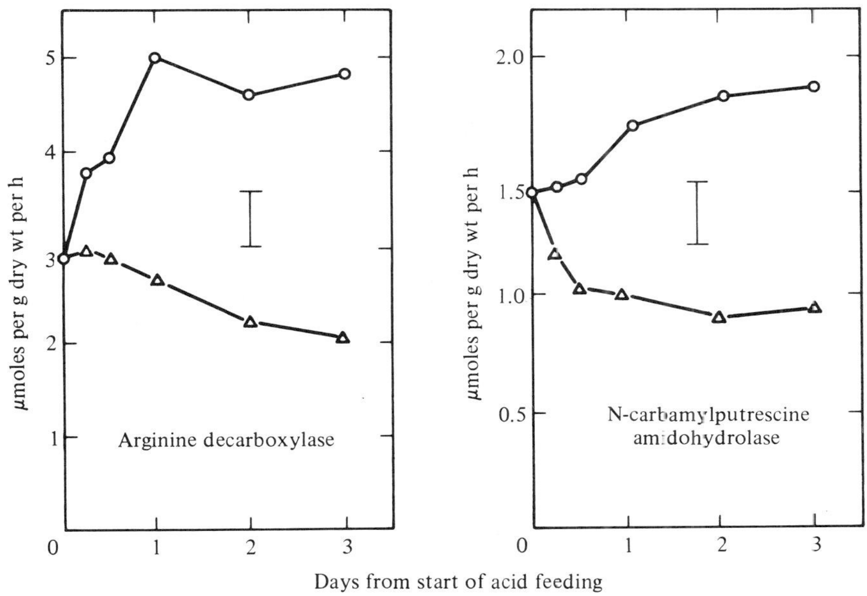

Fig. 67. Arginine decarboxylase and N-carbamylputrescine amidohydrolase activity in the first leaves of barley seedlings fed hydrochloric acid (O) and in the distilled water controls (△) (383).

also significantly increased, but an increased supply of arginine in itself does not evoke the enzyme increases. As the other half of this picture of a homeostatic mechanism at work, the organic acid content of roots falls during K^+ deficiency, and roots and leaves placed in acidic media reduce their content of such acids.

Finally, it may be asked if the production of cadaverine from lysine and the extrusion of the protonated amine in acidic bacterial cultures is not a similar effort to restore intracellular pH via the induced biosynthesis of a lysine decarboxylase of slightly acidic pH optimum (389, 383).

Chapter 4

Organic Cations in the Structure and Function of the Nucleic Acids

---

"A model has been proposed of how the spermine molecule may interact with the deoxyribonucleic acid molecule to stabilize the double-helical conformation of the latter molecule."

MASAMICHI TSUBOI, 1964

We can now realize that our present information on the polyamines, when the studies are actually brought together, comprises an integrated body of knowledge. Its origins stem from the work of such great men as Leeuwenhoek, Vauquelin, Charcot, and Rosenheim. Their discoveries in this area are none the less great for having been largely forgotten. In modern times, basic studies have been made by many capable biochemists and microbiologists, e.g., Snell, Herbst, Mager, the Tabors, Hershey, Ames, Raina, and Williams-Ashman. We can talk about a known physiology and pharmacology of the polyamines, as well as a rigorous classical biochemistry, involving the analysis, biosynthesis, and metabolism of these organic cations. It is clear that the substances are deeply implicated in the most fundamental manifestations of growth and development and are tightly tied in these events to the synthesis of the nucleic acids. The compounds are particularly involved in RNA metabolism and structure, although there are significant intimations that the compounds cannot be neglected in considering the synthesis of DNA viruses or in bacterial division. We are left, however, with many unanswered questions. In particular, what exactly are these substances doing at the molecular level? Therefore, we must now move on to developing a molecular biology of the polyamines. This is easier said than done; nevertheless, such a perspective is required by the present aims and position of our disciplines.

In the last chapter, we did in fact begin our discussion along these lines. We attempted to analyze the role of spermidine in increasing chain initiation and in maintaining chain elongation during synthesis by DNA-dependent RNA polymerase. It was shown that spermidine acted in releasing RNA from the enzyme-template complex at approximate physiological concentrations in a manner similar to a moderately high concentration of Mg^{++} or a fairly high concentration of K^{+}. However, we found it difficult to know the activities or even the concentrations of these cations within cells or at what sites they are combining. Such questions are so difficult to answer that very few workers attempt to answer them, even if they think them important. The dearth of analyses on the cations of the nucleic acids suggests that most workers have not really considered this problem worthy of their attention. But if I have made the importance of these regulatory effects clear, then work is called for in precisely these complicated problems.

Among our psychological difficulties is the desire for specificity. Most workers, when apprised of the data on the activities of the polyamines, will ask if Mg^{++} or a sufficiently high ionic strength will not fulfill roles comparable to those evinced by the polyamines, as in the effects on DNA-dependent RNA polymerase. In many cases this is so, and our colleagues then shrug their shoulders, as if a problem has been answered or has vanished. This attitude seems very well developed among many workers who recognize the existence of multiple paths to the same product or even multiple enzymes for the same reaction but have not yet coped with the possibility of multiple compounds or ions for the same role. One problem is clearly this: if Mg^{++} or spermidine can fulfill similar functions, which one actually operates in the cell in this or that physiological condition? We must ask what determines the cellular choice of the particular alternative actually used. Spermidine can obviously act in some roles that putrescine cannot, and even do better than Mg^{++} in others, as we shall see. Are structural features of the cation and of the recipient anion the determinants in these cases? Frankly, as an approach to these questions, we have little more than a descriptive empiricism to go on.

In our study of the physiology of extreme K^{+} deficiency in bacteria, we observed the striking stimulation of putrescine synthesis but have not yet proceeded to see if the latter effect is elicited at partial K^{+} deficiencies. In the latter instance, the cells are still capable of protein synthesis and growth and the possibility of a physiologically meaningful replacement of a metallic cation by an organic cation may be possible. Similarly, we wonder if putrescine can partially fulfill the role of other polyamines in effects on nucleic acid structure and function. It must be noted that organisms grown in spermine appear to use this unnatural amine to replace putrescine and spermidine. Finally, the existence of homeostatic controls on amine synthesis in plants, and possibly even in bacteria, to control intracellular pH appears to present clear evidence that organisms take the problem seriously even if many scientists do not as yet. It seems that although rigorous specificity has been useful for the evolution of biochemistry, the evolution of organisms may have compelled the selection of alternatives of lesser specificity to cover some consequences of environmental change. Clearly, we are obliged to understand these choices of alternative solutions, as well as the simpler problems with which we may be confronted.

POLYAMINES AND MEMBRANES

I have noted earlier that the polyamines were shown by a variety of workers to protect some fastidious organisms. For example, *Pasteurella tularensis,* which loses viability when washed with distilled water, is protected by the use of saline or water containing spermine (236). Spermine and spermidine at 10^{-3} M will similarly stabilize protoplasts and membranes of *S. faecalis* against osmotic disruption or sonication (148), although not preventing leakage of low molecular components, a distinction not made in earlier studies (237, 406). A similar stabilization of protoplasts by putrescine or cadaverine requires a 10-fold higher concentration. The data suggested that protoplast stabilization is related to the cation-exchange properties of the membrane (310). Both spermidine and spermine will protect *Mycoplasma* strains against lysis (325, 312). It is of interest that polyamines or high concentrations of Mg^{++} (0.2 M) delay lysis of *E. coli* strains infected by ϕX174 or f2 phage, thereby facilitating higher phage yields (137, 139).

Grossowicz and Ariel have reported that spermine and spermidine, but not putrescine or cadaverine, will prevent lysozyme from lysing *M. lysodeikticus* by protecting the membrane and not by inhibiting action of lysozyme (138). On the other hand, the autolytic degradation of the cell wall fraction of a halophilic organism, a reaction that liberates free glycerol, is inhibited by 10^{-3} M spermine (42). Exogenous spermidine will attach itself to isolated cell walls and to protoplasts of *B. subtilis* (12). This result parallels the demonstrated capacity of teichoic acids to bind Mg^{++} ion (149).

The protection of membranes by spermine has been used to stabilize the halophilic membrane-bound enzyme, NADH dehydrogenase, of *Halobacterium cutirubrum* (171). As can be seen in Table 12, the enzyme is not truly salt-requiring since it can be protected by spermine at a relatively low concentration of salt. A similar result has been obtained with the halophilic glycerol dehydrogenase.

Spermine and spermidine produce a marked increase in the metabolism of estradiol by rat liver microsomal membranes. Other aliphatic amines do not have this effect, which is ascribed to a preservation of the lipoprotein structure (193).

Table 12. The Effect of Spermine on the Stability of NADH Dehydrogenase (171)

Incubation conditions	*Specific activity*		*Half-life*
Incubation times:	*0 min*	*120 min*	*(min)*
Experiment I: pH 7.4			
3.4 M NaCl	0.71	0.68	—
0.1 M NaCl	0.61	0	18
0.1 M NaCl + 3.3 mM spermine	0.71	0.45	150
0.1 M NaCl + 33 mM spermine	0.71	0.52	220
Experiment 2: pH 8.5			
3.4 M NaCl	0.66	0.68	—
0.1 M NaCl	0.63	0.03	30
0.1 M NaCl + 3.3 mM spermine	0.68	0.49	200
0.1 M NaCl + 33 mM spermine	0.66	0.63	1500

The enzyme was incubated in 0.1 M Tris hydrochloride at the indicated pH values and cation concentrations. The $-\Delta A_{600m\mu}$ was converted to μmoles NADH, using 19.1 as the millimolar extinction coefficient for dichlorophenolindophenol. Specific activity was defined as μmoles NADH · min^{-1} · mg $protein^{-1}$.

A striking example of this reactivity of spermine with lipids is provided by the work of Shilo (358) on the lipopolysaccharide toxin of the halophilic phytoflagellate that sporadically kills fish farmed in brackish ponds. The potent exotoxin of the alga, *Prymnesium parvum,* attacks gill surfaces and lyses many types of cells. This toxin, which contains sugars, fatty acids, and possibly some phosphorus, is soluble in methanol and in chloroform-methanol but is precipitable by acetone and ether. It is dissociated and inactivated by dialysis although the activity can be somewhat restored by Ca^{++} or Mg^{++} or optimally restored by spermine. We have found spermidine and spermine in *Prymnesium* grown in synthetic medium. It has been shown in gel filtration experiments that spermine will combine with the toxin. As shown in Fig. 68, Shilo proposed that the micellar organization of the anionic toxin is established by the polycationic spermine. Spermine is used routinely in this way to detect a low level of infection of carp pools in Israel (440). If the presence of toxin is detected after activation by spermine, NH_3 is added to the pond. NH_3 is concentrated a thousandfold by the algae, which are thereby killed (359).

Several naturally occurring steroidal diamines, irehdiamine and malouetine, presented in Fig. 69, react with bacterial membranes in such a way that both influx and efflux of K^+ and thiomethylgalactoside (369, 368) are markedly affected. Although steroids themselves are not normally present

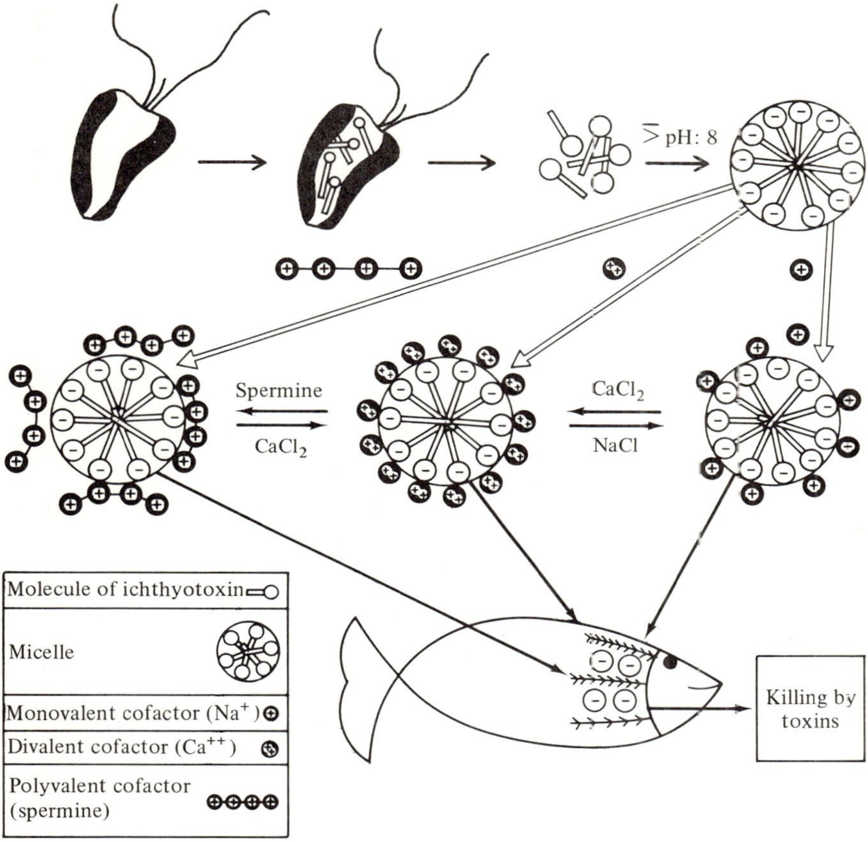

Fig. 68. Mode of action of *Prymnesium* toxin on fish (358)

in *E. coli,* it may be inferred that the lipophilic structure of the diamine facilitates the action of the compounds. Similar effects have been observed with N-acyltriamines (367). The steroidal diamines were also shown by Mahler and his collaborators to combine with DNA (240, 241, 242) and to be mutagenic in phage synthesis and to affect polymer synthesis in *E. coli* (239). The mechanisms of the effects in reactions with DNA are discussed later.

Although these physiological effects on membranes and membrane components are clear enough, we must ask if polyamines can in fact be demonstrated on the membranes in vivo. Quigley and I (300) have isolated the membranes from *E. coli* protoplasts in bulk by means of a zonal ultracentrifuge and have analyzed the membranes for polyamines. Such membranes, derived from an extract degraded by DNase, are low in DNA and

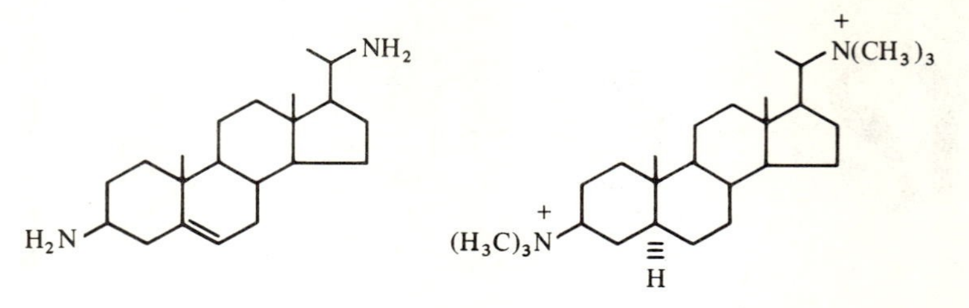

Fig. 69. Structures of two naturally occurring steroidal diamines.

RNA; however, they are relatively concentrated in spermidine, as presented in Table 13. The polyamine is not associated with lipid. To determine the

Table 13. Analyses * of Membrane Preparations of *E. coli* (300)

Sample	*DNA-P mμmoles/cell × 10⁹*	*RNA-P mμmoles/cell × 10⁹*	*Protein μg/cell × 10⁸*	*Putrescine mμmoles/cell × 10¹⁰*	*Spermidine mμmoles/cell × 10¹⁰*	*¹⁴C-Spermidine %*
Initial culture	34.1 ± 1.4	169 ± 5.6	24.2 ± 0.4	130 ± 11.1	63 ± 1.9	100
Mem-brane fraction	0.154 ± 0.049	4.67 ± 0.55	1.59 ± 0.08	1.42 ± 0.37	3.79 ± 1.76	5.9 ± 2.6
%	0.5	2.7	6.6	1.1	6.0	5.9

* Average of 8 experiments.

origin of such spermidine, we mixed membranes from unlabeled cells with the envelope-free lysate from ^{14}C-spermidine-labeled cells and showed that the spermidine on the isolated membrane had completely equilibrated with that in the labeled membrane (300). We are, therefore, unable to say whether any of the spermidine found on the membrane truly belongs there in the intact bacterium or had become associated with the membrane during preparation.

The ease of redistribution of the polyamines on organelles is a perpetual difficulty in the interpretation of studies on the polyamines. Note that the degradation of an *E. coli* extract by DNase does not markedly affect the amount of spermidine found on the membrane. On the other hand, the action of ribonuclease on such an extract produces a significant redistribution of spermidine, a large amount of which becomes sedimentable with the membrane. It can be imagined that spermidine attached to RNA is released by hydrolysis of the nucleic acid for reattachment to other structures.

In many systems the polyamines have striking effects on drug action; these effects may be related to interference with penetration. For example, the substituted diamine, ethambutol, used in treatment of tuberculosis, affects RNA stability in *M. smegmatis* more than the stability of DNA and protein and appears to provoke intracellular degradation of RNA. The inhibitory action of the compound on mycobacteria is reversed by Mg^{++} or spermidine (111). Both together are more effective than either. Despite the initial implication of a direct effect on controls of RNA synthesis, we wonder if the reversing cations do not in part prevent drug penetration. Such a conclusion has been reached concerning the reversing action of spermine toward the action of the schistosomicide, Miracil D, on the intact cells (448). In in vitro systems, on the other hand, this latter compound does combine with DNA and inhibits RNA polymerase. It is possible that spermine also prevents the association of DNA with Miracil D (448, 52).

Some polyamines are known to increase the infectivity of virus nucleic acids in tissue cultures or intact animals. For example, putrescine enhances the infectivity of Mengovirus nucleic acid for mice (58). It is not known if this effect is merely a neutralization of charge on the nucleic acid, or a stabilization of this material, or whether the polyamine affects the membrane itself in some direct way. The problem may be of practical import in the enhancement of induction of interferon (218).

Polyamines and Structures Containing Nucleic Acid

We noted in Chapter 3 that studies such as those by Pardee on polyamines in division or by Hershey and Ames on phage suggest possible roles of polyamines in DNA structure. No one, however, has yet demonstrated the association of naturally occurring polyamine with the bacterial chromosome. It has been observed, nevertheless, that tritiated spermine is incorporated into spheroplasts of *E. coli* and that 60–70% of this spermine is contained in a sedimentable fraction of DNA and RNA observed after spheroplast disruption (196). This labeled spermine is liberated by DNase and not by RNase. Here is a system in which it might be possible to determine if incorporation of the polyamine into the structure containing the nucleic acid requires DNA synthesis.

There is an extensive literature implicating polyamines in ribosome structure. In 1959 Zillig et al. (468) showed that a mixture of free polyamines and an RNA containing bound polyamines result from the dissociation of *E. coli* 70S ribosome by removal of Mg^{++} with EDTA. The polyamine content of the original ribosomes provided one amino group per 3.2 P atoms. Similar fractions were released from liver microsomes.

Polyamines and Ribosomal Structure

At the time of Zillig's report, we were independently studying the properties of ribosomes prepared at high and low Mg^{++} concentrations (66). We also found polyamines in these ribosome preparations and, suspecting that the polyamines might relate to the interaction of ribosomal particles, we studied these effects of exogenous polyamine on aggregation. As described in Chapter 2, we found that 5×10^{-3} M spermidine facilitated the association of 50 S and 30 S subunits to form approximately 70 S particles. Putrescine at comparable levels was ineffective. Since both Mg^{++} and spermidine were more effective in this respect than either alone, we suggested that "Mg^{++} and spermidine are the naturally occurring clasps on the complex ribosomal components existing in the bacterium." In the spermidine-deficient strain of *Pseudomonas,* 10 mM putrescine is partially effective in reassociating ribosomal subunits (264).

E. coli ribosomes may be dissociated by treatment with H_2O_2 and with nitrogen mustard. Such dissociation can be prevented by spermidine, spermine, and divalent metallic cations (277, 263). Putrescine and cadaverine are inactive in preventing dissociation (263). Treated subunits reassociated by polyamines are still capable of synthesis of polyphenylalanine in an appropriate incorporating system. If *E. coli* is grown in the presence of 10^{-4} M spermine, higher ribosomal aggregates are readily obtained (72). Spermine at 0.4 mM associates 30 S and 50 S subunits but at 1 mM precipitates the ribosomes in much larger aggregates (365). With mouse liver ribosomes, which presumably contain spermine already, precipitation begins at 0.1 mM spermine (365).

We also performed experiments, comparable to those described for the membranes, on the redistribution of polyamines during extraction (66, 300). You may recall that exchange was readily detected on membranes (300). However, we observed that the spermidine found on the ribosomes had not been absorbed from the extracting medium after cell distribution. Although this general result has been confirmed by Kim (206) in studies of polyamine exchange on *Pseudomonas* ribosomes, Tabor and Kellogg (263) have observed a greater degree of exchange of spermidine on *E. coli* ribosomes than we have. The basis of this discrepancy is not clear, although binding is reported to be strain dependent. Stevens (399), who has done considerable work on the polyamine content of microsomes as a function of isolation conditions (203), has attributed our result (66) to the low ionic strength at which we worked. Thus, the problem of estimating the true extent of polyamine association with bacterial ribosomes within one cell is not resolved.

It is generally agreed that Mg^{++} at 10^{-2} M in an extracting medium will displace ribosomal polyamines. Just as exogenous Mg^{++} will displace polyamine, the addition of 0.5 mM spermine to pancreatic ribosomes causes the release of ribosomal Mg^{++} and significant amounts of protein, comprising most of the bound secretory proteins, chymotrypsinogen, ribonuclease, and amylase (363). It is quite unclear at this time whether this in vitro reaction observed by Siekevitz and Palade has any functional role, but it should be noted that the pancreas has the highest content of spermine of any mammalian tissue. Polyamines, including spermine, increase production of extracellular enzymes by *B. subtilis* (465).

When leaks in the bacterial membrane occur, it has been found that considerable amounts of K^+ and putrescine are lost together, whereas spermidine is almost quantitatively retained (Fig. 56). Thus, an important association of spermidine with structures within the cell is implied. But which structures? All workers are agreed that ribosomes isolated at relatively low ionic strengths, i.e., conditions that should minimize dissociation and exchange, contain polyamines, which fulfill important structural and functional roles. As mentioned earlier, it has even been reported that *E. coli* grown in low Mg^{++} elaborate more spermidine and that higher concentrations of spermidine are then found on the ribosomes (Table 5).

A certain confusion has arisen by the use of bacterial systems in which the eucaryotic product, spermine, is included in the growth medium. Nutrient broth and yeast extracts contain this substance. Thermophiles or halophiles grown under such conditions have large amounts of spermine bound to ribosomes (400), and it has been asked if this polyamine is not important in protecting the structure in these unusual environments. Spermine has proved to be less readily displaced by acridine orange from thermophile ribosomes than from the ribosomal RNA (400). Despite this suggestion of a structural role of polyamine in the ribosomes of these specialized organisms, displacement of the polyamine by dialysis against M KCl does not completely disrupt these structures.

The eucaryotic *Tetrahymena pyriformis,* although grown in the presence of yeast extract, contains ribosomes that are relatively low in spermine, as can be seen in Table 14. These ribosomes are quite stable in the absence

Table 14. Polyamines and Ribonucleic Acid in Whole Cells and Subcellular Fractions of Tetrahymena (348)

Fraction	RNA-P (μmoles)	Polyamines (μmoles)			Spermidine/ RNA-P	Total poly-amine-N (μmoles)	Molar ratio polyamine-N/RNA-P
		Putrescine	Spermidine	Spermine			
A. Whole cells	13.3	2.73	0.813	0.074	0.061	8.20	0.616
B. Soluble fraction	0.56	5.95	0.612	$<$0.02	—	13.74	24.6
C. Ribosomes *	48.8	2.37	2.13	0.179	0.044	11.85	0.243
D. Ribosomes **	12.3	0.505	0.590	0.057	0.048	3.01	0.245

* Ribosomes prepared by indole method.
** Ribosomes prepared using French press.

of added Mg^{++} and are unusually rich in putrescine and spermine at approximately equimolar concentrations (451). The molar ratio of polyamine N to RNA-P is about 0.24. Perhaps this high content of polyamines in the ribosomes accounts for their unusual stability.

A recent study of the relation of the Mg^{++} and polyamine content of the ribosomes to the function of *E. coli* ribosomes (450) shows that replacement of 70% of the Mg^{++} by polyamine resulted in changes of physical properties of the ribosomes and a precipitous loss of ability to polymerize amino acids. It was suggested that Mg^{++} and spermidine are therefore not functionally equivalent. However, at a certain stage in the removal of Mg^{++}, the loss of some protein did occur (450). It is not possible to conclude, as these workers did, that spermidine was being tested in the mere replacement of displaced Mg^{++}.

It is known that Mg^{++} does not replace a number of other divalent cations (Zn^{++}, Fe^{++}, etc.) in reversing the unfolding of ribosomes caused by EDTA (431), but an activity of spermidine in this respect has not been tested.

As noted earlier, in consideration of the action of polyamines in minimizing degradation of bacterial RNA, spermidine and spermine can protect ribosomes in several ways. *E. coli* ribosomes are rapidly degraded by ribosomally adsorbed ribonuclease in 0.1 M Tris buffer, but such degradation is prevented by 0.1–1 mM spermidine or spermine (351). Similar effects have been observed with the ribosomes of brain cortex; in addition, the polyamines increase the melting temperature of the ribosomes (79).

The Reaggregation of Dissociated Plant Viruses

Spermidine has been found in some plant viruses. Indeed, spermidine can account for about 20% of the cation associated with the RNA in turnip yellow mosaic virus (29). Recognition of some role of the polyamines in ribosomal organization has led to studies of their possible effects in the degradation and aggregation of plant viruses and their self-assembly. In studies on tobacco mosaic virus by Brakke and Van Pelt (39), it was observed that Mg^{++} or polyamine as low as 10^{-4} M inhibit degradation of the particles. On the other hand, concentrations of 10^{-2} M stimulate aggregation. EDTA and citrate eliminate these effects of Mg^{++} but not of the polyamines.

Cowpea chlorotic mottle virus normally contains spermidine. As shown by Hiebert and Bancroft (160), the virus can be disassembled in M NaCl, 0.02 M Tris, and 10^{-3} M Clelands reagent. The virus can then be reassembled in a similar medium in which M NaCl is replaced by 0.01 M KCl and 5×10^{-3} M Mg^{++}, as shown in Fig. 70. The replacement of Mg^{++} by

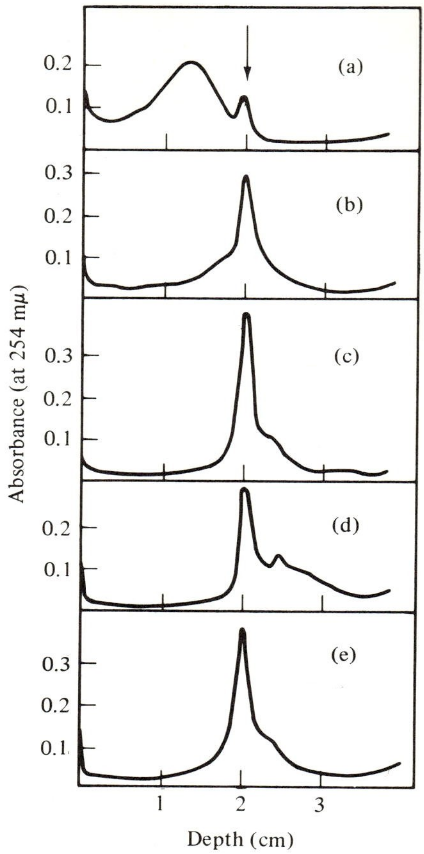

Fig. 70. Sedimentation distribution of reassembled cowpea chlorotic mottle virus (CCMV) (160). (a) The absence of divalent cation; (b) the presence of 5×10^{-3} M $MgCl_2$; (c) spermidine; (d) spermine; and (e) 3,3′-diaminodipropylamine. Centrifugation was for 50 min at 69,000 rpm in a SW 65 rotor with gradients of 100–400 mg of sucrose per ml of 0.1 M pH 5.7 acetate buffer. The arrow marks the position reached by CCMV.

spermidine or spermine is also presented in the figure. The product reassembled in spermidine had an infectivity comparable to the Mg^{++} control. The reassembled virus particles behave in many physical respects like normal virus. Comparable results were obtained with two other spherical viruses. The fact that assembly occurs rapidly and correctly at low temperatures and ionic strengths in the presence of divalent cations implies the major role of ionic interactions in this assembly.

However, reassembly is far less precise at pH 5.7. This result has suggested that the divalent cations are reacting with protein as well as with virus RNA structure which would be less affected at pH 5.7. It is therefore believed that configurations assumed by the proteins in the presence of the divalent cations are also crucial in the reassembly. However, the analysis has not advanced to the point of proving these suppositions.

POLYAMINES, RIBOSOMES, AND PROTEIN SYNTHESIS

The apparent interchangeability of Mg^{++} and spermidine in maintaining ribosome structure has permitted tests of the essentiality of these cations in polypeptide synthesis. A number of workers, of whom Nathans and Lipmann (279) were the first, have shown that the transfer of ^{14}C-aminoacyl-tRNA to *E. coli* ribosomes can operate to a considerable extent with spermidine at low Mg^{++} concentrations. In *Salmonella typhimurium* (356), *B. subtilis* (163), yeast (40), or mammalian systems (157), the polyamines were soon seen to stimulate such transfers at suboptimal Mg^{++} concentrations. Typical experiments with *S. typhimurium* preparations are given in Fig. 71. In Fig. 71(a) the suboptimal concentration of 12 mM Mg^{++} was used, whereas Fig. 71(b) shows that the optimal Mg^{++} concentration in this system was 17 mM.

Martin and Ames concluded that polyamines appear to act in this transfer reaction primarily by increasing the fraction of aggregated ribosomal subunits (250). This view is not contradicted by many of the studies in the literature on such systems. However, the presence of the polyamines does increase particle stability in several additional ways, e.g., by increasing the Tm of ribosomes and of ribosomal RNA and decreasing degradation by contaminating or added nucleases (103, 79). For example, it is stated that mescaline-induced breakdown of ribosomes isolated from drug-treated brain slices can be prevented by the presence of spermine (79). We shall see later that the polyamines affect steps in protein synthesis other than that of associating ribosomal subunits.

In the *Pseudomonas* strain lacking spermidine, 10 mM putrescine facilitates aggregation of ribosomal subunits and polyphenylalanine synthesis. Dialysis of the ribosomes of this strain reduces diamine to fewer than 20 molecules of diamine per 70 S ribosome (258). However, 20 mM Mg^{++}, which is relatively high, permits full activity in polyphenylalanine synthesis from phenylalanyl-tRNA. Thus, the bulk of ribosomal polyamines do not serve an obligatory function in this system. The optimal Mg^{++} concentration can be partially replaced by spermidine, but the polyamines will not permit synthesis with such ribosomes in the complete absence of Mg^{++} (435).

Despite the absence of an apparently obligatory role of polyamine in

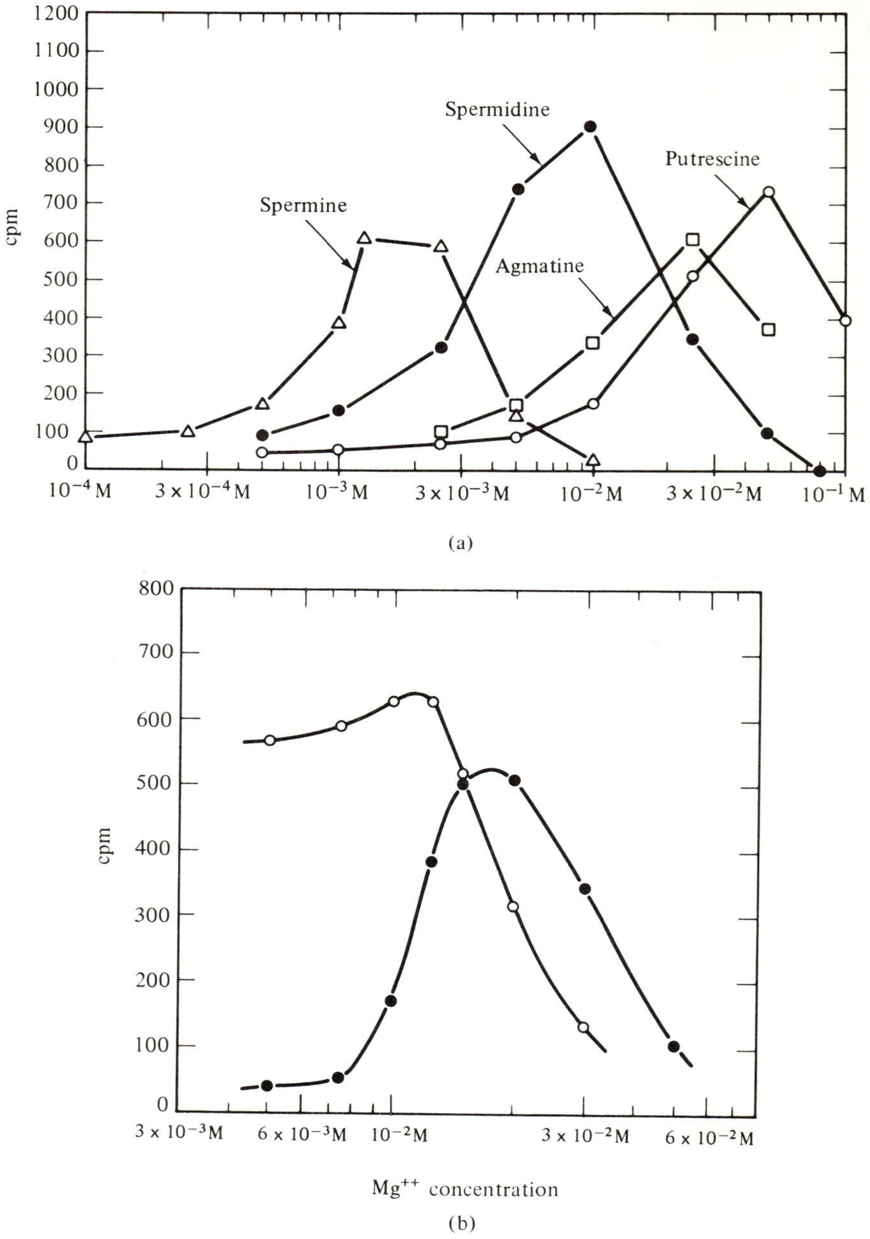

Fig. 71. (a) Stimulation by polyamines of poly U-directed phenylalanine incorporation in *S. typhimurium* ribosomes (250). (b) Phenylalanine incorporation as a function of Mg^{++} concentration: (●), no polyamines; (O) 5 mM spermidine and 25 mM putrescine added to the assay mixture (250).

many systems,[9] the data of Deutscher et al. (84) on systems derived from *B. megaterium* vegetative cells or spores suggest the major specific role that can be played by spermidine at suboptimal Mg^{++}. This is shown in Fig. 72. Table 15 shows that 2 mM spermidine has a relatively enormous effect on

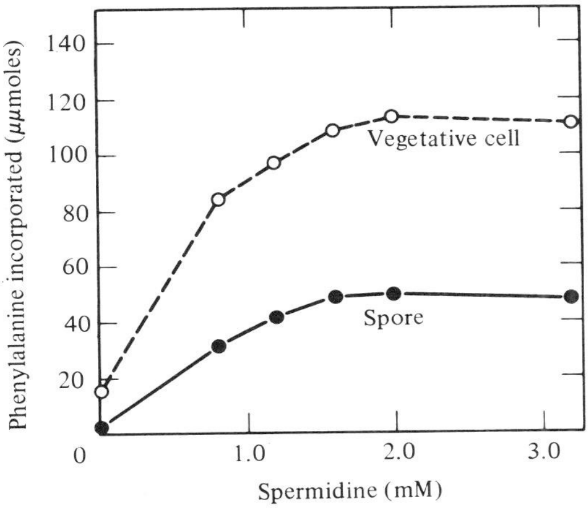

Fig. 72. Stimulation by spermidine of poly U-directed phenylalanine incorporation in *B. megaterium* ribosomes (84). Incubations were carried out at 12 mM Mg^{++}.

the system and can produce at least twice as much incorporation at 6 mM Mg^{++} as with 12 mM Mg^{++}. Optimal Mg^{++} was at 15 mM in this system.

The data in Table 15 also show that other polyamines, particularly spermine above 10^{-4} M, can be markedly inhibitory in the system. Similar observations and study by Ochoa and Weinstein in a mammalian system have suggested that spermine may also tie up mRNA and tRNA (284). In addition, Ochoa and Weinstein have reported that the polyamines can affect the fidelity of translation of synthetic polynucleotides (121, 449).

Other Sites of Polyamine Action in in Vitro Protein Synthesis

Results comparable to those presented for *B. megaterium* were obtained some years earlier by Mager with rat liver (157) and by Halvorson with yeast systems (155). These workers concluded that the spermidine effect was greater than a "mere" replacement of Mg^{++}. Halvorson found, for ex-

[9] It has been reported most recently that the amino acid-incorporating system derived from *Bacillus thuringiensis* has an obligate requirement for spermidine in addition to magnesium ions (55). This requirement is not due to an inhibition of nucleases; the substance appears to be essential for the activity of the *B. thuringiensis* ribosomes.

Table 15. Effect of Polyamines on Poly U-Directed Phenylalanine Incorporation (84)

Source of ribosomes	*Polyamine added*	*Phenylalanine incorporated*	
		6 mM Mg^{++}	*12 mM Mg^{++}*
		μμmoles	
Vegetative cells	None	1.0	99.2
	Spermidine	300	154
	Spermine	15.4	12.2
	Putrescine	6.2	117
	Cadaverine	0	113
Spores	None	0	77.2
	Spermidine	353	142
	Spermine	6.5	10.6
	Putrescine	0.5	124
	Cadaverine	0	114

Incubations were performed in Reaction Mixture B lacking spermidine except for the addition of either 6 mM or 12 mM Mg^{++}. *E. coli* tRNA (200 μg) and poly U (80 μg) were added to each reaction mixture. Spore S_{100} fraction, 270 μg, was added to each incubation mixture, together with 70 μg of either vegetative cell or spore ribosomal protein. Polyamines, when present, were at 2 mM.

ample, that the transfer to polypeptide from phenylalanyl-tRNA could be significantly stimulated even at optimal Mg^{++} by 1 mM spermidine (155). In pursuing this transfer in a yeast system, Tanner (432) has shown that spermidine also stimulates the binding of yeast tyrosyl tRNA to washed ribosomes in the presence and absence of messenger RNA or of appropriate trinucleotides (Fig. 73). In the yeast system, spermine is optimally active at 0.5 mM, i.e., about 1/50 of the spermidine requirement, whereas putrescine is inhibitory at all concentrations. Binding at all concentrations of Mg^{++} is markedly stimulated by a 50:1 mixture of spermidine and spermine. As Tanner has indicated, the polyamines are not merely substituting for Mg^{++} in the yeast systems.

Yeast ribosomes isolated from vigorously growing yeast do not dissociate readily. Tanner has suggested that the different activities of yeast ribosomes isolated from different stages of the life cycle may, in fact, relate to differences in their polyamine content, a point that has not been proved.

Takeda has studied the interchangeability of Mg^{++} and polyamine in *E. coli* systems synthesizing polyphenylalanine (426) and MS2 phage-directed polypeptides (427). He has also demonstrated that in this system

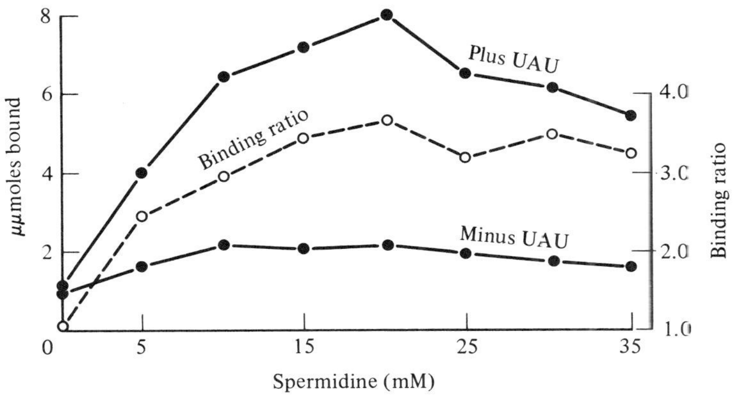

Fig. 73. Effect of spermidine on binding of yeast [^{14}C] tyrosyl-tRNA to yeast ribosomes. The binding ratio is ratio of moles of tyrosyl-tRNA bound in the presence of mRNA to moles bound in its absence (432).

spermidine can replace Mg^{++} in the binding of aminoacyl tRNA to ribosomes (428).

Furthermore, Takeda reported that the polyamines stimulate aminoacyl tRNA formation in the absence and presence of Mg^{++} (429, 430). It had been shown earlier in Cantoni's laboratory (34) that spermidine serves as well as Mg^{++} in the transfer of amino acid from a ^{14}C-seryladenylate-enzyme complex to tRNA. This is exemplified in Table 16. Takeda's data suggested that a spermine-tRNA complex can be acylated without added

Table 16. Metal Ion Requirement for Transfer Step (373)

Metal	*^{14}C-serine transferred*
	cpm
None	44
Mg^{++}	763
Mn^{++}	640
Cd^{++}	713
Ca^{++}	917
NH_4^+	84
Cu^{++}	18
$Cu^{++} + Mg^{++}$	17
Spermidine	760

Transfer assay as described under "Methods" (373) except that the magnesium was replaced by 5 μmoles of the various metal ions as indicated, and 1300 cpm ^{14}C-seryladenylate-enzyme complex were used per incubation mixture. The chloride salt of all metals was used.

Mg^{++} (430, 176). He showed that spermine binds to tRNA but not to the enzyme, even in the presence of ATP. Moreover, he was able to establish conditions in which the degree of aminoacylation was proportional to spermine in the system (176).

Several antibiotics have been observed to interfere with the binding of aminoacyl tRNA to the 30 S ribosomal subunit at low Mg^{++} concentration. These include pactamycin, streptomycin, and chloramphenicol. This inhibition of binding is reversed by 10 mM spermidine or 20 mM Mg^{++} (60, 59). It is possible that this result accounts for the observation in 1962 by Mager et al. (238) showing that Mg^{++} or spermidine can prevent streptomycin inhibition of leucine incorporation in a bacterial amino acid-incorporating system.

Polyamines and Some Other Enzymes

The effects of polyamines on numerous enzyme activities were reviewed in 1964 (421). Since that time, enzymology has developed a considerable interest in quaternary structure, and this phenomenon appears to be particularly relevant to the problems of the role of the polyamines.

In the case of uridine diphosphate galactose-4-epimerase of yeast, three proteins of 3.6 S, 6.5 S, and 10.9 S have been detected, of which the first is inactive (77, 78). Although association of two 3.6 S subunits to a 6.5 S particle is thought to be essential to activation, further association to the 10.9 S particle is probably not as important. Activation by cations, in which the polyamines are particularly effective, is frequently observed to involve an increase of sedimentation rate of epimerase from 6.5 S to 10.9 S, as shown in Fig. 74. However, spermine gives a half maximal stimulation

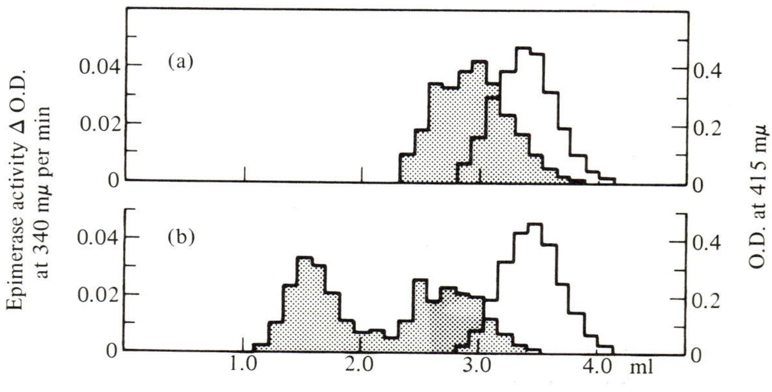

Fig. 74. Effect of 5 × 10^{-3} M spermine on the sedimentation of UDP-galactose-4-epimerase (77) in a sucrose density gradient. (a) Control gradient, no spermine; (b) gradient containing 5 × 10^{-3} M spermine. The stippled area represents epimerase activity. The solid line represents OD at 415 mμ (hemoglobin).

of the enzyme at 2×10^{-4} M, manifested as a marked lowering of the Km of the enzyme for UDP-galactose (78), and at about this concentration of spermine no 10.9 S component is detected. In this instance, it is not clear that activation by spermine, which is real enough, actually occurs by increasing subunit association.

The conversion of glycogen phosphorylase b by adenylic acid (AMP) into phosphorylase a results in the incorporation of two moles of phosphate per mole of phosphorylase b, with the concomitant appearance of catalytic activity and subunit association. Although AMP increases the aggregation of b, subunit association is greatly enhanced by the polyamines, which do not produce such an association without AMP (445). The polyamines appear to increase the affinity of phosphorylase for AMP but not for glucose 1-phosphate. The results shown in Fig. 75 indicate that in the presence of AMP, the polyamine also facilitates the aggregation of monomers to an active tetramer. We have recently analyzed a preparation of crystalline phosphorylase a isolated from rabbit muscle and have detected significant amounts of spermidine and spermine in this preparation.

Similar activations by spermidine and other cations have been observed by Sanwal for the allosteric phosphoenolpyruvate carboxylase of *Salmonella typhimurium* (346) and for the glucose 6-phosphate dehydrogenase of *Escherichia coli* (345). As mentioned earlier, in the latter case the activation may facilitate the production of ribonucleotide precursors essential for increased RNA synthesis. Thus, we may soon be in the position of adding aggregation effectors to appropriate enzyme systems. Perhaps we do this unconsciously in adding protamine or streptomycin to eliminate nucleic acids from extracts. As a result, I wonder whether the aggregation of RNA polymerase subunits in the presence or absence of the DNA template would be facilitated by the polyamines and account in part for activation of this enzyme.

Extraction of bacteria in spermidine-containing buffers lacking Mg^{++} not only aggregated ribosomal subunits but also facilitated the sedimentation of cellular and viral DNA (66), as well as much mRNA, at low speed with the bacterial cell walls and membranes (64). This result has led us to ask if the polyamines do not normally help to organize a hypothetical intracellular site at which the bacterial chromosome, membrane, and the nucleic acid polymerases are centered. Driedger (89a) has recently shown by electron microscopy that exposure of cells to spermidine before fixation with osmium tetroxide appears to reveal multiple attachment sites between DNA and membrane.

The extraction with spermidine instead of Mg^{++} minimized the presence of nucleotide precursors generated via Mg^{++}-activated polynucleotide phosphorylase. The absence of Mg^{++} also prevented the action of DNase and Mg^{++}-activated kinases. Thus, the presence of spermidine permitted a more

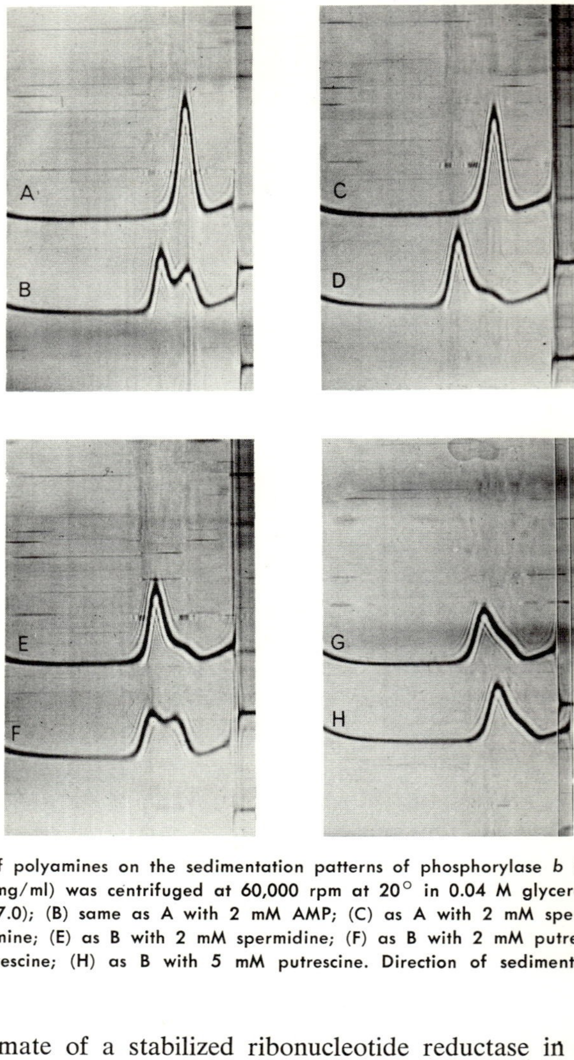

Fig. 75. Effect of polyamines on the sedimentation patterns of phosphorylase *b* (445). (A) Phosphorylase *b* (5 mg/ml) was centrifuged at 60,000 rpm at 20° in 0.04 M glycerophosphate-0.03 M cysteine (pH 7.0); (B) same as A with 2 mM AMP; (C) as A with 2 mM spermine; (D) as B with 2 mM spermine; (E) as B with 2 mM spermidine; (F) as B with 2 mM putrescine; (G) as B with 3 mM putrescine; (H) as B with 5 mM putrescine. Direction of sedimentation was from right to left.

accurate estimate of a stabilized ribonucleotide reductase in the extracts (64). Reichard has shown that the active association of the *E. coli* enzyme subunits requires divalent cations (315). Can polyamine substitute for such cations in this association? Beck (28) has stated that the divalent cation requirement of the reductase of the *Lactobacillus leichmanii* system can be replaced by spermidine. In any case, it is evident that numerous problems of practical enzymology might be eased by a judicious use of the polyamine.

Another area that has been increasingly studied during the past decade has been that of the role of lipids in a number of enzymes and in organelle structure. Razin and Rozansky (314) described the precipitation of phospholipids (lecithin) by spermine and other polyamines and suggested that this reaction, as well as reactions with nucleic acids, could account for the binding of polyamines to cells, protoplasts, membranes, etc. It would be of interest in this connection to study the reaction of polyamines with several purified enzymes whose structures and activities are known to depend on the presence of phospholipids. The report of Salganicoff (344) on the stimulation by polyamines of succinate-linked respiration in mitochondria and on the maintenance of oxidative phosphorylation in aging or Ca^{++}-treated organelles is particularly relevant.

Some Reactions of the Polyamines with DNA

The discovery of polyamines in the T-even phages, which contain DNA, led to many studies on the stabilization of phage and phage nucleic acids by the polyamines. As mentioned earlier, Tabor described such stabilization of T5 by spermidine (416). Fraser and Mahler had observed that urea-damaged phage was stabilized in this way by polymethylene diamines, of which cadaverine is the most effective (115). This soon led them to study also protection by polyamines against proflavine in dark and light inactivation (116). Cadaverine was maximally protective against dark inactivation. This group was also among the first to show that the diamines, of which diaminopentane was the most effective, increased the melting temperature of DNA (244, 243) and of helical polyribonucleotides (255). The Tabors demonstrated independently such effects by polyamines; first on transforming DNA (417), then on calf thymus DNA (418). They eventually performed a detailed study with Kaiser of the protection by spermine against shear of λ DNA (198). Some of these results are shown in Fig. 76. Higher diamines, e.g., diaminononane, have been shown to protect DNA against denaturation by X-irradiation (43). Putrescine actually accelerates denaturation of this type (43).

The results of Stevens (398) on the increase of T_m of DNA in a dilute buffer as a function of the concentration of various polyamines indicate that spermidine and spermine are far more active than is putrescine. As presented in Fig. 77, a 3-2-3 tetramine is much less active than spermine or spermidine molecules in which the central portion is longer than two methylene groups. Charge is clearly not the sole consideration, although a non-specific effect of a cation shielding phosphates must also partly account for the data.

If the T_m of DNA-polyamine complexes is studied at different concentrations of Na^+ (173), a competition is obtained between Na^+ and polyamine NH_3^+ for the nucleotide phosphate, as in Fig. 78. Two processes are

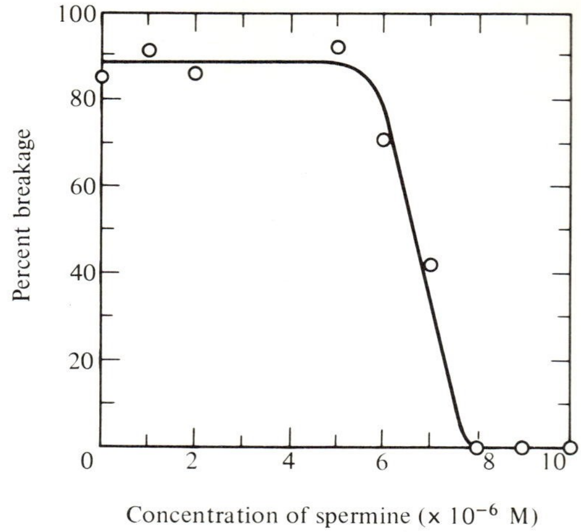

Fig. 76. The DNA of coliphage λ at 2.6 × 10^{-6} M nucleotide-P in 10^{-2} M Tris (pH 7.1) -10^{-3} M EDTA was stirred at 3600 rpm for 30 min in a 5 ml beaker containing 4 ml of solution. Spermine was added less than 4 min before stirring was begun. The midpoint of the drop in fragility occurs where there are 10 spermine-N atoms per DNA-P (198).

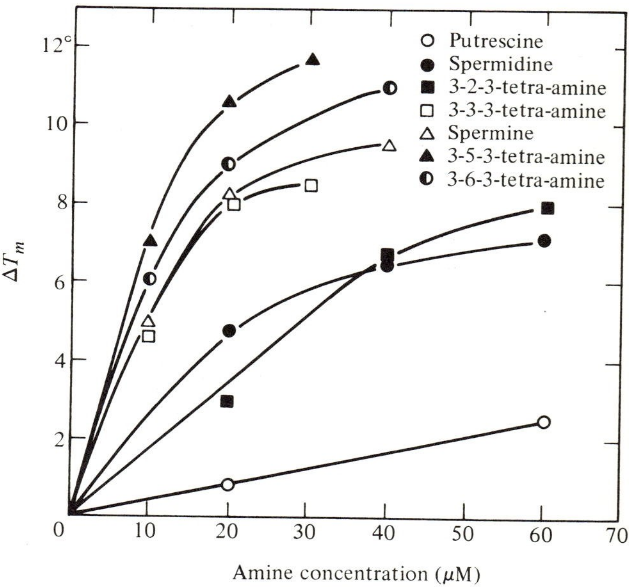

Fig. 77. Change of T_m of calf thymus DNA with varying concentrations of added amines (398). The DNA was initially dissolved in 0.03 M NaCl-1 mM potassium dimethylglutarate buffer, pH 6.2, at a final concentration of 60 μg/ml. The T_m of the DNA in the absence of added amine was 73.5°.

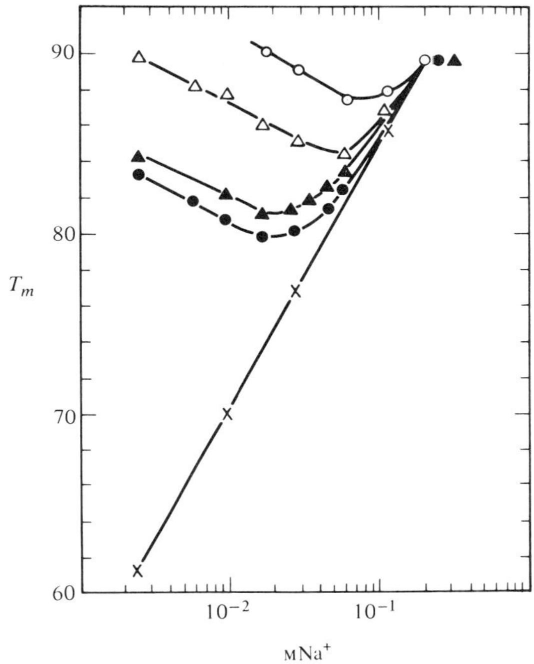

Fig. 78. Dependence of T_m of calf thymus DNA in the presence of polyamines on Na^+ concentration in 2.2×10^{-4} M citrate (173). Na^+ was added as NaCl. △, 8.05×10^{-5} M spermidine · 3 HCl; ●, 2.35×10^{-4} M spermidine · 3 HCl; ▲, 2.35×10^{-4} M hexanediamine · 2 HCl; ○, 2.35×10^{-4} M putrescine · 2 HCl.

operating as a function of increasing Na^+: destabilization due to dissociation of the DNA-polyamine and stabilization with increasing ionic strength.

In 1964 Tsuboi (438) proposed that both primary and secondary amino groups of spermine interact with four DNA phosphates, with the secondary amino groups making a bridge across the shallow groove in the double helix while the primary amino groups were bound to the adjacent nucleoside phosphate along a single polynucleotide chain. Tsuboi suggested that, in forming an oblique bridge across the shallow groove between the two polynucleotide helices, the polyamine may function as a clamp for binding two chains together. Essentially this model has been adopted by Liquori (231), who has also presented some X-ray evidence to suggest that spermine and spermidine can fit as proposed. Such models are presented in Fig. 79. The data of Stevens (398) on the relative inefficacy of the 3-2-3 analog to increase T_m would seem to support this model. Suwalsky et al. (402a) have interpreted their recent X-ray study of a spermine-DNA complex to suggest that the polyamine can stabilize the naturally occurring B form of DNA in solution. The data on their complex were best interpreted as a combina-

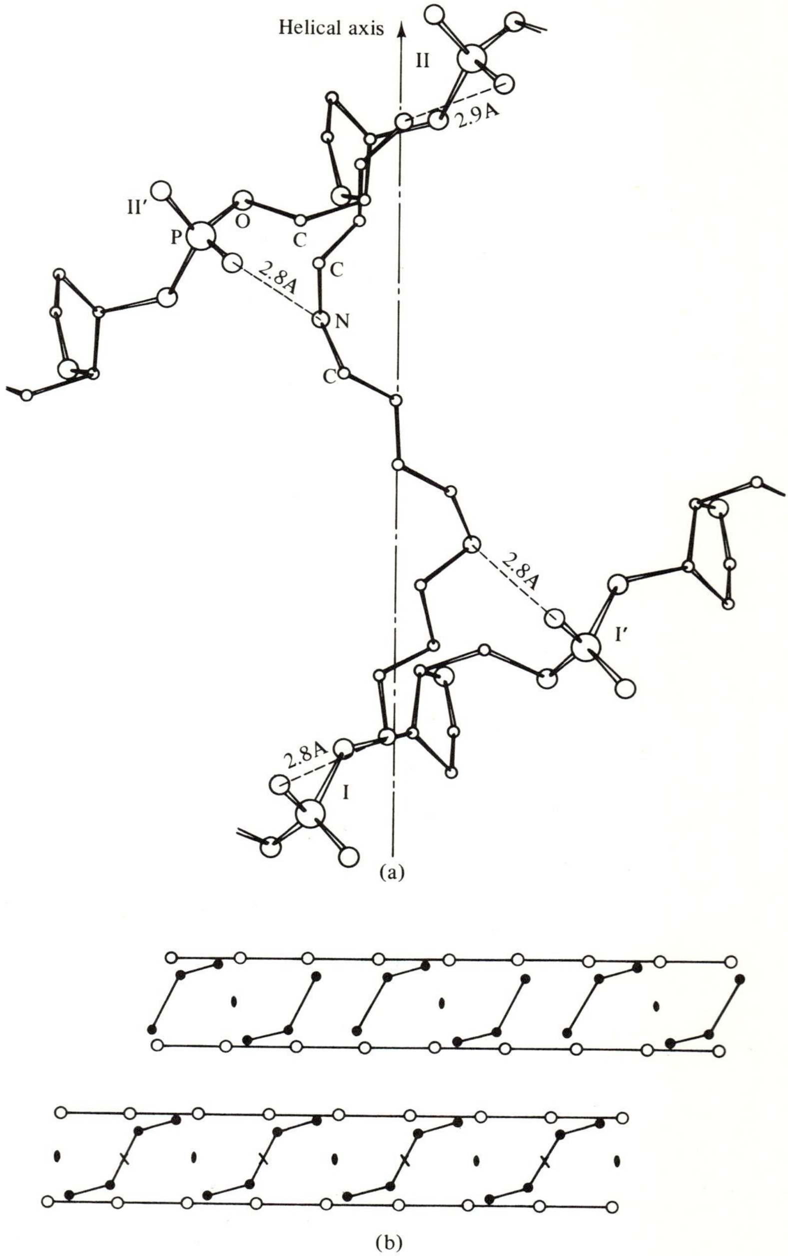

Fig. 79. (a) Detailed scheme of the complex between DNA and spermine projected on a cylindrical surface. (b) Models showing spermidine and spermine molecules in the narrow grooves of DNA (231).

tion of the binding of spermine across the small DNA groove in one molecule plus intermolecular crosslinks from DNA to DNA.

Both Tsuboi (43) and Liquori et al. (231) have suggested that, in complexing with DNA, spermidine tends to stabilize the ordered double-stranded structure in solution. On the other hand, the insertion of the carcinogen, 3,4-benzpyrene, into DNA is presumed to depend on the existence of defective regions in this structure. Liquori and his colleagues have shown that spermidine reduces the solubility of benzpyrene in DNA (230), presumably by stabilizing and reducing the number of defective regions in the DNA structure.

Studies on the binding of various substituted diamines to native double-stranded DNA have attempted to test the Tsuboi-Liquori model. In this model stabilization is effected through links across the narrow groove, rather than the stabilization possibly occurring through neutralization of repulsive charges on the separate polynucleotide chains (132). Steroidal diamines comprised of diprimary amines, such as irehdiamine (Fig. 69) destabilize the double-stranded structure at a ratio of amine/DNA-P > 0.25. On the other hand, the substituted steroidal diamine, malouetine, stabilizes the structure of dA-dT regions preferentially. Although it is recognized that bulky substituted amines prevent a steroid such as the stabilizing malouetine from substituting for a base pair, the labilization by irehdiamine is not thought to arise from intercalation (240, 241, 242). In the absence of viscometry, this evidence is not conclusive at present concerning the problem of penetrating the DNA groove.

Studies of Gabbay et al. on the interaction of "reporter" molecules, i.e., nitroaniline-labeled polyamines (124), with DNA have suggested strongly that the reporter molecules are bound to the minor groove of the helix. Investigations by proton magnetic resonance, viscometry, and other methods have indicated that the nitroaniline-labeled polyamines do, in fact, intercalate between base pairs in DNA (289, 123). Such evidence thus tends to support the Tsuboi-Liquori model.

In a recent paper on tryptamine and harmine, it was concluded that these compounds bind to DNA largely by intercalation (384). As mentioned in Chapter 2, the antimalarial agent, chloroquine (Fig. 23), is believed to have its 7-chloroquinoline ring intercalate between base pairs of DNA after it is positioned across the minor groove by ionic interaction of its substituted putrescine side chain with appropriate phosphates. The effects of chloroquine binding on the viscosity, sedimentation, and optical dichroism of DNA were consistent with this model (283).

Although various groups have observed that the effect of spermine on the T_m of various DNAs appears to change with base composition (247), Felsenfeld found that the net charge of spermine varies with temperature in such a way as to indicate that the T_m changes may relate to the temperature used to melt the DNA samples. The results of studies by equilibrium dialysis (Fig. 80) show that there is no significant variation either in the number of strongly binding sites or in the strength of binding with base composition. It was concluded that there is no evidence of correlation between binding of spermine and the base composition of DNA (170). This non-specific binding of spermine has been used in studies of dialysis competition to show that some ions, e.g., tetramethylammonium ions, do have selectivities for DNA of particular base compositions (356).

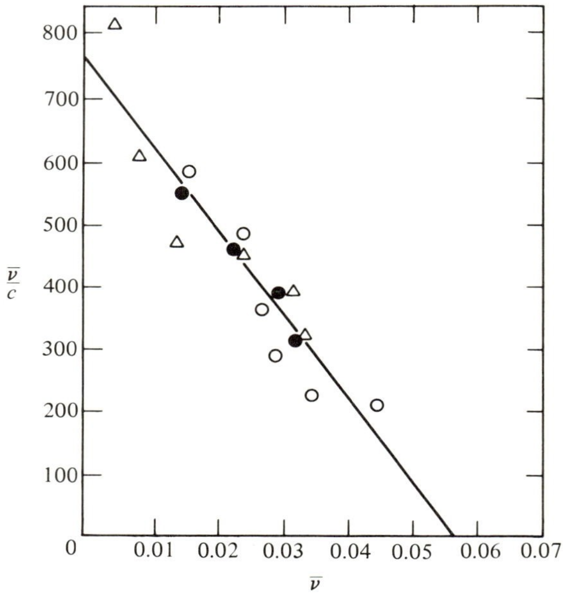

Equilibrium Constant K and Number of Sites N for Binding of Spermine to Native DNA at 4°C

DNA source	A-T	$K \times 10^{-4}$	N
Proteus mirabilis	62	1.31 ± 0.09	0.058 ± 0.002
Calf thymus	58	1.40 ± 0.26	0.055 ± 0.005
Serratia marcescens	42	1.35 ± 0.30	0.058 ± 0.006

Fig. 80. Results of equilibrium dialysis experiments. $\bar{\nu}$ is the number of moles of spermine bound per mole of nucleotide; c is the concentration, in moles/liter, of free spermine; (O) calf thymus DNA; (●) *Proteus mirabilis* DNA; (△) *Serratia marcescens* DNA (170).

Reactions of Polyamines with Polyribonucleotides

Spermidine and spermine stabilize double-stranded and triple-stranded helical complexes of polyadenylate (poly A) and polyuridylate (poly U), as well as helical complexes of polyinosinate and polycytidylate (132, 252, 162). The three-stranded (A + 2U) molecules are stabilized by spermine more than is the two-stranded (A + U). Furthermore, the effect of the polyamine is markedly stronger than that of Mg^{++}. Nevertheless, the higher valence of spermine could well account for the marked affinity of the polyamine for the multistranded structure, so that the postulated "clamping" function of spermine need not be invoked.

On the other hand, poly U and polyribothymidylate (poly riboT) also show surprising increases in T_m at equivalent concentrations of nucleotide and counterion (405, 404, 403), as shown in Fig. 81. These results are

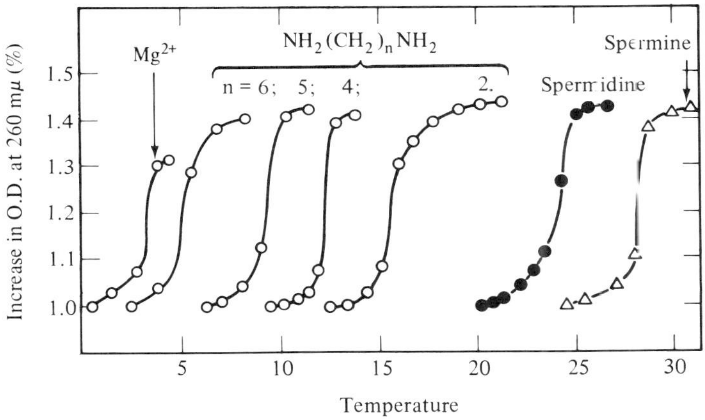

Fig. 81. Melting of poly U in the presence of Mg^{2+} and amines (405). Polymer concentration at 4×10^{-5} M, concentration of divalent cations 10^{-3} M. Concentration of polyamines was equivalent to polymer phosphate.

unlike those with poly C and poly A whose melting profiles are not affected by polyamines (404). It had been thought that poly U exists as a random coil at room temperature (272, 178), although at 0°, Mg^{++} ion confers an enhanced hypochromicity. The addition of polyamine below the T_m produces an immediate drop in absorbance. The transitions in the system are completely reversible; because they occur over a sufficiently narrow temperature range, a cooperative process is indicated. Thus, the polyamines appear to markedly increase secondary structure of these polynucleotides. Studies of the optical rotatory dispersion of poly U-spermine complexes have revealed a very high degree of base stacking (326). The suggestion that the poly U-polyamine structure might be double-stranded (326) has not been sustained in enzymatic test (347). Poly U-polyamine complexes, as compared to normal poly U, do have markedly altered responses to ultraviolet light (81).

Poly U and poly C are rendered more resistant to pancreatic ribonuclease by formation of the spermine complexes, although no one has proved that the development of helical structures themselves impede hydrolysis (200). Poly A also becomes significantly more resistant to ribonuclease in the presence of spermine or spermidine (125). The latter results have been interpreted to mean that in this structure, the polyamine binds to adjacent phosphate anions and that the complexed regions will have lower affinities for the active site of the enzyme.

When we turn to natural RNA, the situation obviously becomes more complex than with homopolymers. The melting temperature of yeast RNA is increased by spermine (134). Similarly, the RNA of turnip yellow mosaic virus is converted to a more stable and compact structure by polyamine (262). Such a structure shows a higher T_m and S rate and a lower reduced viscosity; it is also more stable to ribonuclease.

Cantoni reported in 1960 (50) that tRNA can be fractionated with spermine, whereas Felsenfeld and Huang (107) showed the considerable affinity of polyamine for tRNA and various homopolymers by conductimetric titration. More recently, Muench (273, 276, 274, 275) has demonstrated that the substituted diaminobutane, chloroquine, specifically activates the tryptophanyl tRNA of *E. coli.* Almost simultaneously, Fresco (117) demonstrated an activation of tRNA by spermidine. We shall consider the possible significance of these effects shortly.

The ability of polyamines to interact with polymeric anions is obviously not limited to the nucleic acids. They react well with various acidic polysaccharides, such as heparin, chondroitin sulfate, etc., compounds that may on occasion contaminate nucleic acid preparations (4). In addition to observing the evident need for caution in binding experiments with natural products, we note that the interaction of the polyamines with such substances raises many questions concerning the possible significance of such reactions in vivo.

Polyamines and tRNA

I shall bring my discussion to a close with some data on the interaction of polyamines with tRNA. The various facts implicating the polyamines in the transcription of DNA and in protein synthesis are listed in Fig. 82. It can be seen that there are at least nine different phenomena in this metabolic sequence in which the polyamines are involved; the category of effects related to tRNA structure may be further subdivided.

In considering all of these possible candidates for an experimental analysis of the molecular role of polyamines in nucleic acid structure and function, we have been led to the study of tRNA, whose structure is known to be responsive to polyamines. This type of RNA is a significant fraction

DNA		
↓	Transcription	1. Strand selection 2. Chain initiation 3. Chain extension 4. Chain termination
RNA		
↓	Translation	5. Preparation of tRNA 6. Aminoacylation of tRNA 7. Message binding to 30 S subunit 8. Aminoacyl tRNA binding to 30 S subunit 9. Ribosomal subunit association
Protein		

Fig. 82. Reactions in the sequence DNA→RNA→protein in which the polyamines have been implicated.

of cellular RNA and possesses a considerable degree of structural homogeneity among its numerous species. The nucleotide sequences of many species of *E. coli* and other tRNA are completely known (298); the various types of molecules can cocrystallize (119, 31); and, as a result of crystallographic and other analyses, the three-dimensional structure is well on the way to complete definition (466, 217, 74, 224). Transfer RNA is made in bacteria under relaxed conditions and relatively simple methods have been devised for its isolation. Although these isolation methods were unsuitable for our purposes, it seemed possible to modify them in order to make them suitable. Since tRNA structure reacts to polyamine, we tried to see if tRNA could be isolated from cells with polyamine attached. Furthermore, we studied the effects of varying cellular metabolism on the polyamine content of such tRNA.

As listed on Table 17, the polyamines have been shown to affect tRNA in at least four types of reaction. Several groups of workers have found that spermidine or spermine can replace Mg^{++} or other divalent cations in facilitating crystallization of tRNA (119, 147). It has been observed and described by Fresco (118, 225–228), Muench (273–276), and Sueoka (118, 225–228, 128, 180–182) that tRNA may exist in an inactive configuration. This configuration can be converted to an active one by high concentrations of monovalent cations (1–2 M) or by low concentrations (about 1 mM) of Mg^{++} or of spermidine (117). The inactive and active forms are separable on Sephadex gels or on methylated albumin-kieselguhr.

Table 17. Effects of Spermidine on tRNA

1. Crystallization		
2. Inactive	$\overset{Mg^{++} \text{ or spd}}{\rightleftharpoons}$	Active configuration
3. tRNA + SAM	$\xrightarrow[\text{Liver methylase}]{Mg^{++} \text{ or spd}}$	CH_3-tRNA
4. E·AMP·AA + tRNA	$\xrightarrow{Mg^{++} \text{ or spd}}$	tRNA·AA

Leboy has described the activation of a partially purified liver tRNA methylase by Mg^{++} or by polyamine (221). In this instance, the polyamines appear to be better activators than Mg^{++} in producing a greater extent of methylation, as seen in Table 18. A comparison of the rates of reaction in the presence of the polyamines is seen in Fig. 83.

Table 18. Extent of Methylation of RNA Substrates[a] (221)

	mμmoles CH_3 incorporated/mg RNA				
Substrate	*No additions*	*MgAcetate (10 mM)*	*Spermine (0.08 mM)*	*Spermidine (0.5 mM)*	*Putrescine (10 mM)*
Rat liver sRNA	—	< 0.4	< 0.4	< 0.4	< 0.4
E. coli K12 methyl-deficient sRNA	< 0.4	5.3	18.6	20.6	20.2
E. coli K12 normal sRNA	—	2.8	8.7	9.4	9.6

[a] Each reaction tube contained 4 μg sRNA and EDTA-washed ammonium sulfate extract equivalent to 1 mg protein in a final volume of 0.25 ml. [Methyl-^{14}C]S-adenosylmethionine had an activity of 65 cpm/μμmole; therefore, < 0.4 mμmoles CH_3/mg RNA indicates less than 100 cpm incorporated/4 μg sRNA.

Finally, as noted earlier, it was found (34) that the transfer of an amino acid from an isolated enzyme-amino acid-adenylate complex to acylate tRNA requires Mg^{++} or spermidine and proceeds equally well with either cation. Furthermore, the most recent study (176) implicates a polyamine-tRNA complex as the active acceptor of the amino acid. It there-

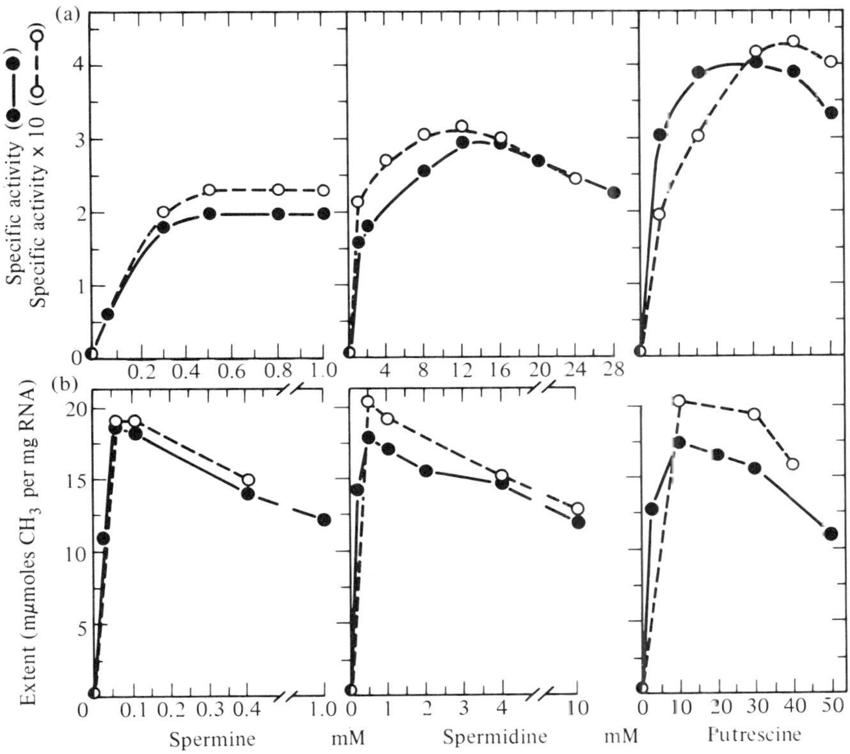

Fig. 83. Effect of polyamines on partially purified rat liver methylase (221). (a) Specific activity; (b) extent of methylation; (O - - - O) EDTA-treated ammonium sulfate fraction; (●——●) DEAE-Sephadex peak.

fore appears possible that samples of tRNA containing polyamine can be found in which the enzymatic reactions could proceed without added Mg^{++} or polyamine.

The Isolation and Identification of tRNA

Published procedures for the purification of tRNA have involved fractionations, e.g., elution from ion exchangers, at high ionic strength. These procedures might be expected to dissociate polyamines from the nucleates. Actually, our analyses have revealed small but occasionally significant amounts of organic cations in preparations of tRNA from eucaryotic cells, which contain the less dissociable spermine. However, 13 different homogeneous preparations of *E. coli* tRNA sent to us from the Oak Ridge Laboratory contain much less than 0.05 mole of spermidine or putrescine per molecule of tRNA.

We have devised an isolation procedure in which the tRNA is maintained at low ionic strength (68). The RNA is released from relatively small batches of frozen cells by phenol and extracted into an aqueous phase, resulting in a rapid reduction of salt concentration. The crude extracted nucleic acid is precipitated at 67% ethanol from 0.025 M K acetate (pH 7.0) and contains about 4 OD units at 260 mμ per mg, of which about 40–60% is tRNA. The preparations contain cell wall components and small amounts of ribosomal RNA and 5 S RNA.

The crude nucleates were dissolved in 0.01 M K acetate (pH 7) and were fractionated at 4° on a column of Sephadex G-100, as described in Fig. 84(a). In most preparations three peaks were obtained, as can be seen in the figure. With the nucleates from *E. coli* strain 15 TAU, the initial fractions immediately following the void volume contain a small but significant amount of RNA, presumed to be rRNA, in addition to some cell wall material. This initial fraction does not accept amino acids in an activating system.

The initial shoulder to the main peak apparently contains material approaching 5 S RNA. Subsequently this fraction was clearly resolved on a longer column; moreover, it did not accept amino acids. The central three-quarters of the main peak accepted amino acids (phenylalanine and others) in an amount proportional to optical density. The accepting quality of this fraction per unit of optical density is comparable to or better than that of *E. coli* tRNA purified on a methylated albumin-kieselguhr column or a variety of commercial preparations. Fractions in this peak also accept methyl from S-adenosylmethionine in the presence of a liver tRNA methylase (221) and accept serine from the complex of seryladenylate-activating enzyme (34).

The third small peak appears to consist of relatively low molecular metabolic products.

Analysis of Polyamines in tRNA

These studies became possible as a result of the adaptation (68, 87) of the analytical procedure introduced by Seiler and Wiechmann (353). This method can measure polyamine in the range of 10^{-10} to 10^{-11} moles per acidified samples and permits the estimation of the polyamine contained in nucleates of 0.01 optical density unit at 260 mμ. Amino acids do not interfere since their dansyl derivatives do not move in the chromatography in the solvent used. Spermine is readily detected on the same chromatogram as spermidine and putrescine.

We have spent a great deal of our efforts studying tRNA after it has been isolated by the above method from a pair of isogenic stringent and

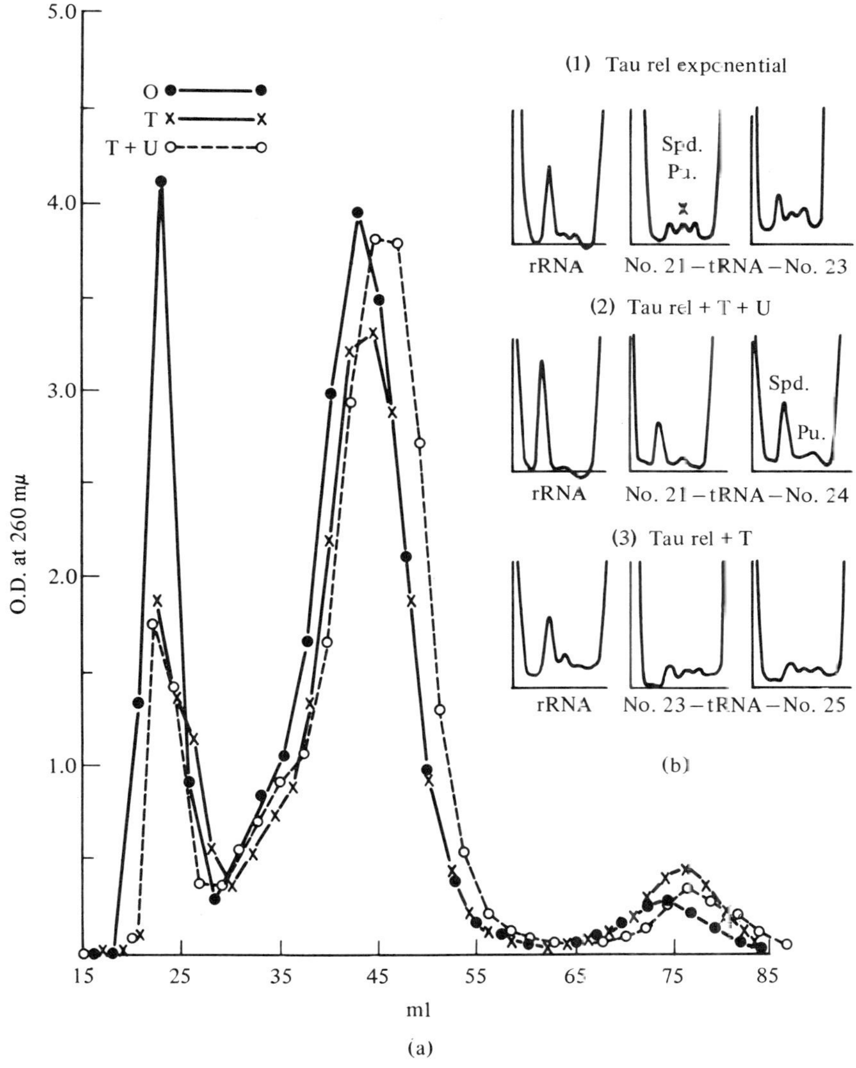

Fig. 84. (a) The purification of tRNA on Sephadex G100 in 0.01 *M* potassium acetate at pH 7.0 (68). Three different samples of crude tRNA from TAU rel were fractionated on the same column at 4° on successive days. The O sample was derived from organisms harvested during exponential growth. Aliquots of this same culture were washed and incubated further for 1 hr in the absence of arginine in the mineral medium plus glucose supplemented with thymine (T) or with thymine plus uracil (T + U). (b) Tracings of dansylated polyamines of RNA fractions from TAU rel (68) in exponential growth (1), or incubation in thymine plus uracil (2), or thymine alone (3).

relaxed mutants, *E. coli* strain 15 TAU. In strains of this organism, which require thymine, uracil, and arginine for growth, polyamine can be synthesized in the absence of arginine. This organism is blocked between ornithine and citrulline and produces putrescine from ornithine. Although the absence of thymine does not markedly affect polyamine synthesis, it does provoke thymineless death, and we routinely include thymine in the cultures. RNA synthesis can be followed by measuring uracil incorporation. RNA synthesis in strain 15 TAU can be varied in many ways, such as by using the stringent or relaxed strains or by adding antibiotics such as chloramphenicol to stimulate synthesis of RNA in the stringent organism. As noted earlier, in all of these systems, expanded RNA synthesis involves an increase in spermidine. Furthermore, the relaxed mutant makes far more spermidine than does the stringent cell in the presence or absence of RNA accumulation.

In examination of the RNA fractions from the relaxed cell harvested during exponential growth, we found that both the presumed ribosomal RNA and tRNA contain spermidine. The ribosomal RNA fraction is richer in polyamines than is the tRNA [Fig. 84(b)]. Since the fractionation takes place on a buffered Sephadex column, it is clear that the spermidine found in the RNA fractions was not exchangeable with the K^+ ion contained in a dilute buffer. Both RNA fractions were fairly low in putrescine, but this cation is much higher than is spermidine in *E. coli.* Although the rRNA fraction contained about 3 molecules of spermidine per 77 nucleotides (229), the tRNA was heterogeneous, containing for example 0.7 moles of spermidine per mole of tRNA on one side of the peak and 1.5 moles per mole on the other.

Part of the culture, harvested during exponential growth, was permitted to make RNA for 1 hr in the absence of arginine. The crude tRNA was isolated and fractionated. The analyses of the fractions showed an increase of spermidine in "rRNA" to 9 molecules per 77 nucleotides, whereas the spermidine in tRNA had increased slightly but significantly. The tRNA now appeared homogeneous with respect to spermidine content at a level of about 2 molecules per 77 nucleotides, or 2 moles per mole of tRNA.

We also analyzed the tRNA from another portion of the same culture exposed to thymine without arginine or uracil. In the absence of uracil, no accumulation of RNA or of spermidine was detected although spermidine continued to be synthesized. It is excreted from the cell as acetylspermidine. We might have imagined that newly synthesized spermidine was added to preformed tRNA. Nevertheless, the analysis of the fractions revealed a pattern similar to that observed in the exponential cell. One possible interpretation of this result is that spermidine is added to the structure of tRNA soon after biosynthesis in the course of its folding to form an active molecule.

When we compared the RNA fractions from a stringent cell harvested in exponential growth, we observed different patterns. The ribosomal fraction was rich not only in spermidine but also in an unknown dansyl derivative. This unknown component was present in some tRNA fractions but not in others. Unlike the rRNA, the tRNA fractions of the stringent cells were relatively low in spermidine. Thus, there is some specificity of the nucleic acids in determining the distribution of these organic cations.

The tRNA fractions, both low and high in spermidine, have been tested in their ability to accept methyl from S-adenosylmethionine or serine from a seryladenylate-enzyme complex in the absence of added spermidine or Mg^{++}. In both cases, additional Mg^{++} was necessary to permit the reactions to occur even in fractions relatively high in spermidine. Similar studies are being continued.

We have observed that the tRNA is elaborated in stringent TAU stimulated by chloramphenicol. In polyamine content, such RNA is much like the tRNA produced in relaxed TAU.

We have also examined the tRNA isolated from a stringent strain of *E. coli* unable to concentrate K^+. When this bacterium is placed in Na^+, there is a relaxed synthesis of rRNA and tRNA. As we noted in Chapter 3, the cells in Na^+ reduce spermidine production by 75% but are stimulated to make a very large amount of putrescine. The tRNA made under these conditions is relatively low in spermidine but does contain significantly more putrescine than do our other isolates of tRNA. Thus, the absence of putrescine in other tRNA isolates suggests that only the polyamine available during tRNA synthesis is added to the structure in a relatively nondissociable form.

We have examined the tRNA isolated from mouse fibroblasts in exponential and stationary phases of growth. The fractionation presented in Fig. 85 reveals a larger proportion of rRNA than in the fractionation of bacterial RNA. Of greater interest, however, are the polyamine contents of the preparations presented in the figure. Both ribosomal and tRNA contain spermine as well as spermidine and lack putrescine. The spermidine to spermine ratios in both exponential growth and stationary phases of the cells were about 3:1, although the concentration of both substances had almost doubled on entering the stationary phase. It can be seen that the relative contents differ significantly in the two types of rRNA. Apparently, the tRNA from exponential cells may contain more spermidine than spermine, whereas the reverse is true in the tRNA from stationary cells. In this comparison, it appears that the polyamines on tRNA need not reflect the gross composition of polyamines in the cell.

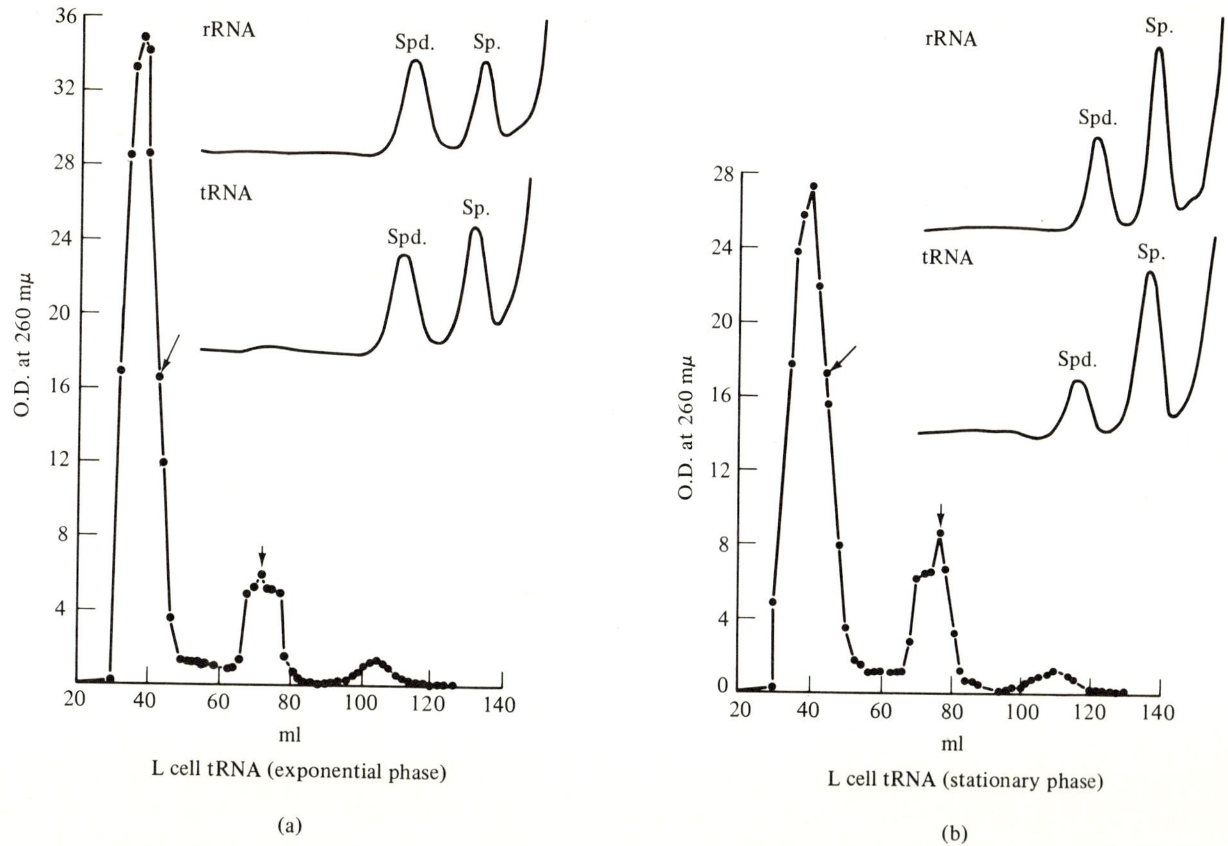

Fig. 85. Fractionation of RNA of mouse fibroblasts (L cells) grown in tissue cultures and tracings of dansylated polyamines (63a). sp = spermine; spd = spermidine; (a) exponential growth; (b) stationary phase.

Does Polyamine Exist on tRNA Within the Organism?

Exponential stringent and relaxed TAU have similar putrescine and spermidine contents. Nevertheless, the differences between the nucleic acid fractions of the two strains seem to suggest that their polyamine contents do not reflect merely the gross polyamine composition of the cell.

The experiments on the tRNA of TAU in various conditions indicated that the distribution of polyamine on the tRNA fraction can be quite dissimilar from that on rRNA. Thus, the differences among nucleic acids themselves appear to contribute to the polyamine content observed.

In all of these instances, the relative absence of putrescine on these fractions may suggest merely the much greater affinity of spermidine to nucleic acid. However, when less spermidine is available, as in the K^+-deficient bacterium incubated in Na^+, the tRNA fractions bind relatively more putrescine. The fact that the amount of polyamine including putrescine in any tRNA fraction does not exceed 2–3 moles total per mole, despite the excess of polyamine, suggests that the sites binding polyamine possess some unusual properties.

We have begun to correlate the RNA isolated with newly synthesized RNA detected by incorporation of radioactive uracil. Since the incorporation of uracil indicates that only a fraction of the existing RNA (about 15%) is newly made, the completely altered pattern and the apparent homogeneity of the fractions with respect to polyamine may suggest a fairly rapid and extensive turnover (or exchange) of polyamine in the molecules.

As a result of the difficulty of performing truly critical control experiments in this system, we are unable to answer as yet the basic question posed by the presence of polyamine on isolated tRNA. We can isolate tRNA with attached polyamine from bacteria by a relatively mild procedure, but we are not yet certain that the attached polyamine is not an artifact of the preparation procedure. Nevertheless, the fact that spermidine activates tRNA in several defined reactions and that it is found on tRNA in low stoichiometric amounts indicates that the structural relationship of the polyamines to tRNA should be clarified more precisely. We should know the nature of the binding sites on tRNA for the polyamine and whether they are homogeneous or heterogeneous. If there are several types of binding sites, we should attempt to clarify the nature of the structures that can bind spermidine in a relatively nondissociable form.

The Possible Structural and Functional Significance of the Polyamines in the Nucleic Acids

As noted earlier, Liquori and his colleagues presented a model (Fig. 79) for the binding of spermidine and spermine molecules in the narrow grooves of DNA. This model, involving relatively close-fitting three-point and four-point attachments of polyamines to the phosphates of the polynucleotide chains, could account fairly well for the observed nondissociability of spermidine-DNA complexes. These complexes do not appear to dissociate even during electrophoresis (173).

Such a model may also be applicable to the organization of helical loops in single-stranded RNA. Thus, the well-known tRNA open cloverleaf [Fig. 86(a)] is seen as held together in some areas by as few as three complementary base pairs. We could believe that this limited association in these areas is not sufficient to maintain the structure. However, it is now accepted that tRNA is a folded structure, and various models are offered for the three-dimensional structure. The crystallographic studies of the Wisconsin group (466, 217) have suggested one such model [Fig. 86(b)]. An examination of their folded structure reveals that the limited helical regions defined in the cloverleaf structure also appear relatively unsupported in the new model. It can be suggested that spermidine may form a lock in such limited helical regions to prevent the unraveling of the structure.

If polyamines facilitate looping and the maintenance of limited double-stranded helices in single-stranded RNA, they can be protecting such RNA against nucleases. The preservation of loops and helices is desirable in the economy of the cell rRNA and tRNA. We can imagine that the evolution of short double-stranded loops locked by polyamine is an easier accomplishment than the evolution of large numbers of base pairs. In addition, perhaps the same mechanism will eliminate the translation of much of the single-stranded RNA of the cell. Ribosomal RNA, for example, can be translated if opened by denaturation in the presence of bulky cations like neomycin. It is also possible that the same mechanism may serve to punctuate certain polycistronic virus RNA molecules (R17 bacteriophage RNA) whose polynucleotide structures have been found to contain sufficient numbers of complementary base pairs to permit the formation of loops (3). In addition, such base-paired loops may facilitate the packaging of certain RNA messages, like turnip yellow mosaic virus RNA in the virion. That there are problems of preservation and non-translation of certain RNA molecules seems clear. Although it is not yet proved that the polyamines in nature participate in the solution of these problems, they should be considered as candidates for both roles.

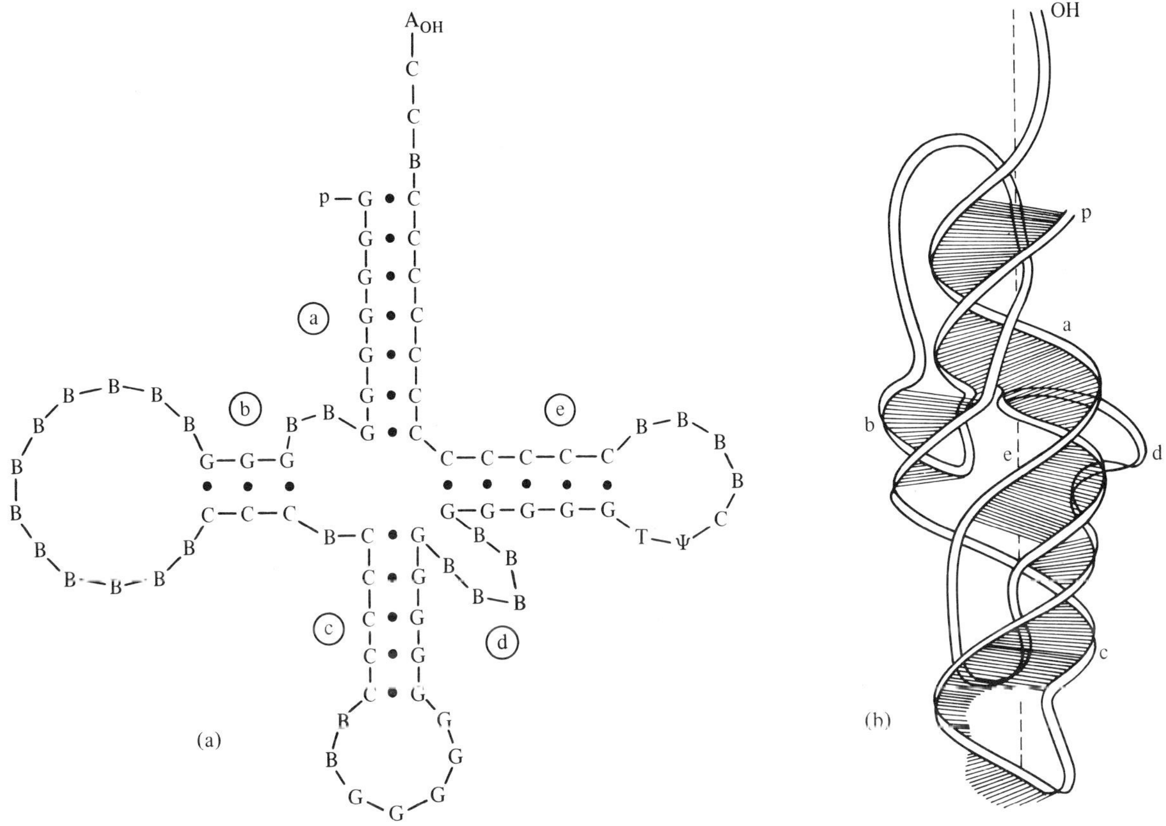

Fig. 86. Cloverleaf configuration of average tRNA. The B is an average unpaired base. The CCA terminus and TΨC loop are identified. The bottom loop contains the anticodon. The five unpaired but helically stacked bases in this loop were treated as guanine. The large dihydrouridine loop is on the left (217). (b) Schematic drawing of folded model labeled as in (a) (217). The dashed line is the long helix axis.

Conclusion

I have described in some detail all the reasons I believe we must be concerned with the polyamines. These substances are present in all cells in large amounts and appear to possess the potentiality of substituting for cations such as K^+ or Mg^{++} that may be present in short supply. They react with many compounds and help to organize structure in cells, organelles, and individual polymers. They control division and growth and play roles in almost every step in the synthesis of RNA and protein. Obviously, they seem to be involved in too many activities; for this reason, I expect you to be suspicious. However, there are good reasons to suppose that you should look into these matters anyway. For one thing, it is refreshing to think about a little known but potentially exciting field. For another, it would be just too unfortunate to make the mistake of not helping to clarify an important area in which there is so much to be done. Finally, we may suggest to the cautious ones that the footsteps of Leeuwenhoek, Vauquelin, Charcot, and Rosenheim can be followed with a moderate measure of security.

Bibliography

1. Abraham, K. A. 1968. European J. Biochem. *5*:143.
2. Achee, F. M., C. H. Chervenka, R. A. Smith, and K. T. Yasunobu. 1968. Biochemistry *7*:4329.
3. Adams, J. M., P. G. N. Jeppesen, F. Sanger, and B. G. Barrell. 1969. Nature *223*:1009.
4. Agrell, I. P. S., and O. Heby. 1968. Exp. Cell Res. *50*:668.
5. Alarcon, R. A. 1970. Arch. Biochem. Biophys. *137*:365.
6. Alarcon, R. A., G. E. Foley, and E. J. Modest. 1961. Arch. Biochem. Biophys. *94*:540.
7. Ames, B. N., and D. T. Dubin. 1960. J. Biol. Chem. *235*:769.
8. Ames, B. N., D. T. Dubin, and S. M. Rosenthal. 1958. Science *127*:814.
9. Anderson, N. G., and C. B. Norris. 1960. Exp. Cell Res. *19*:605.
10. Audit, C., B. Viala, and Y. Robin. 1967. Comp. Biochem. Physiol. *22*:775.
11. Bachrach, U., S. Abzug, and A. Bekierkunst. 1967. Biochim. Biophys. Acta *134*:174.
12. Bachrach, U., and I. Cohen. 1961. J. Gen. Microbiol. *26*:1.
13. Bachrach, U., and G. Eilon. 1967. Biochim. Biophys. Acta *145*:418.
14. Bachrach, U., and G. Eilon. 1969. Biochim. Biophys. Acta *179*:473.
15. Bachrach, U., and G. Eilon. 1969. Biochim. Biophys. Acta *179*:494.
16. Bachrach, U., and J. Leibovici. 1965. Biochem. Biophys. Res. Comm. *19*:357.
17. Bachrach, U., and J. Leibovici. 1966. J. Mol. Biol. *19*:120.
18. Bachrach, U., and I. S. Oser. 1963. J. Biol. Chem. *238*:2098.
19. Bachrach, U., and S. Persky. 1966. Biochem. Biophys. Res. Comm. *24*:135.
20. Bachrach, U., and S. Persky. 1969. Biochim. Biophys. Acta *179*:484.
21. Bachrach, U., S. Persky, and S. Razin. 1960. Biochem. J. *76*:306.

22. Bachrach, U., S. Rabina, G. Loebenstein, and G. Eilon. 1965. Nature *208*:1095.

23. Bachrach, U., and B. Reches. 1966. Anal. Biochem. *17*:38.

24. Bachrach, U., C. W. Tabor, and H. Tabor. 1963. Biochim. Biophys. Acta *78*:768.

25. Bachrach, U., and A. Weinstein. 1970. J. Gen. Microbiol. *60*:159.

26. Barros, C., and G. Giudice. 1968. Exp. Cell Res. *50*:671.

27. Bases, R., and F. Mendez. 1969. J. Cell. Physiol. *74*:283.

28. Beck, W. S. 1967. J. Biol. Chem. *242*:3148.

29. Beer, S. V., and T. Kosuge. 1970. Virology *40*:930.

30. Berlin, R. D. 1969. Arch. Biochem. Biophys. *134*:120.

31. Blake, R. D., J. R. Fresco, and R. Langridge. 1970. Nature *225*:32.

32. Blaschko, H. 1968. Nature *219*:1164.

33. Blethen, S. L., E. A. Boeker, and E. E. Snell. 1968. J. Biol. Chem. *243*: 1671.

34. Bluestein, H. G., C. C. Allende, J. E. Allende, and G. L. Cantoni. 1968. J. Biol. Chem. *243*:4603.

35. Boeker, E. A., E. H. Fischer, and E. E. Snell. 1969. J. Biol. Chem. *244*:5239.

36. Boeker, E. A., and E. E. Snell. 1968. J. Biol. Chem. *243*:1678.

37. Borek, E., E. J. Rockenbach, and A. Ryan. 1956. J. Bact. *71*:318.

38. Boyle, S. M., and P. S. Cohen. 1969. J. Bact. *96*:1266.

39. Brakke, M. K., and N. Van Pelt. 1969. Virology *39*:516.

40. Bretthauer, R. K., L. Marcus, J. Chaloupka, H. O. Halvorson, and R. M. Bock. 1963. Biochemistry *2*:1079.

41. Brewer, E. N., and H. P. Rusch. 1966. Biochem. Biophys. Res. Commun. *25*:579.

42. Brown, A. D. 1960. Biochim. Biophys. Acta *44*:178.

43. Brown, P. E. 1968. Radiat. Res. *34*:24.

44. Busch, H., K. Shankar Narayan, and J. Hamilton. 1967. Exp. Cell Res. *47*:329.

45. Caldarera, C. M., B. Barbiroli, and G. Moruzzi. 1965. Biochem. J. *97*:84.

46. Caldarera, C. M., P. P. Giorgi, and A. Casti. 1969. Boll. Della Soc. Ital. Biol. Speri. *45*:460.

47. Caldarera, C. M., M. S. Moruzzi, B. Barbiroli, and G. Moruzzi. 1968. Biochem. Biophys. Res. Commun. *33*:266.

48. Caldarera, C. M., M. S. Moruzzi, C. Rossoni, and B. Barbiroli. 1969. J. Neurochem. *16*:309.

49. Campello, A. P., C. W. Tabor, and H. Tabor. 1965. Biochem. Biophys. Res. Commun. *19*:6.

50. Cantoni, G. L. 1960. Nature *188*:300.

51. Cantoni, G. L., and J. Durrell. 1957. J. Biol. Chem. *225*:1033.

52. Carchman, R., I. B. Weinstein, and E. Hirschberg. 1968. Bact. Proc. 32.

53. Cashel, M., and B. Kalbacher. 1970. J. Biol. Chem. *245*:2309.

54. Cedarleaf, C., and S. S. Cohen. 1971. J. Bact. in press.

55. Changchien, L., and J. N. Aronson. 1970. J. Bact. *103*:734.

56. Chargaff, E., M. Ziff, and S. S. Cohen. 1940. J. Biol. Chem. *136*:257.

57. Church, R. B., and B. J. McCarthy. 1967. J. Mol. Biol. *23*:459.

58. Cobet, A. B., and T. G. Akers. 1968. Appl. Microbiol. *16*:1257.

59. Cohen, L. B., I. H. Goldberg, and A. E. Herner. 1969. Biochemistry *8*:1327.

60. Cohen, L. B., A. E. Herner, and I. H. Goldberg. 1969. Biochemistry *8*:1312.

60a. Cohen, S., B. W. O'Malley, and M. Stastny. 1970. Science, *170*:336.

61. Cohen, S. S. 1942. J. Biol. Chem. *144*:353.

62. Cohen, S. S. 1945. J. Biol. Chem. *158*:255.

63. Cohen, S. S. 1968. *Virus-Induced Enzymes*. Columbia Univ. Press, New York.

63a. Cohen, S. S. 1970. Ann. N. Y. Acad. Sci. *171*:169.

64. Cohen, S. S., and H. D. Barner. 1962. J. Biol. Chem. *237*:PC1376.

65. Cohen, S. S., N. Hoffner, M. Jansen, M. Moore, and A. Raina. 1967. Proc. Nat. Acad. Sci. *57*:721.

66. Cohen, S. S., and J. Lichtenstein. 1960. J. Biol. Chem. *235*:2112.

67. Cohen, S. S., and J. Lichtenstein. 1960. J. Biol. Chem. *235*:PC55.

68. Cohen, S. S., S. Morgan, and E. Streibel. 1969. Proc. Nat. Acad. Sci. *64*:669.

69. Cohen, S. S., and A. Raina. 1967. *In* H. J. Vogel, J. O. Lampen, and V. Bryson [ed.] *Organizational Biosynthesis,* p. 157, Academic Press, New York.

70. Cohen, S. S., and W. M. Stanley. 1942. J. Biol. Chem. *142*:863.

71. Cohen, S. S., and W. M. Stanley 1942. J. Biol. Chem. *144*:589.

72. Colbourn, J. L., B. H. Witherspoon, and E. J. Herbst. 1961. Biochim. Biophys. Acta *49*:422.

73. Coleman, R. G., and F. J. Richards. 1956. Ann. Bot. N.S. *20*:393.

73a. Couse, N. L., D. J. Cummings, V. A. Chapman, and S. S. DeLong. 1970. Virology *42*:590.

74. Cramer, F., V. A. Erdmann, F. von der Haar, and E. Schlimme. 1969. J. Cell. Physiol. *74*(suppl.):163.

75. Crocomo, O. J., L. C. Basso, and O. G. Brasil. 1970. Phytochemistry *9*:1487.

76. Danzig, M., and H. P. Schultz. 1952. J. Am. Chem. Soc. *74*:1836.

77. Darrow, R. A., and R. Rodstrom. 1966. Proc. Nat. Acad. Sci. *55*:205.

78. Darrow, R. A., and R. Rodstrom. 1970. J. Biol. Chem. *245*:2036.

79. Datta, R. K., S. Sen, and J. J. Ghosh. 1969. Biochem. J. *114*:847.

80. Davidson, D., and N. G. Anderson. 1960. Exp. Cell Res. *20*:610.

81. De Boer, G., M. Pearson, and H. E. Johns. 1967. J. Mol. Biol. *27*:131.

82. De Marco, C., and B. Mondovi. 1960. Ital. J. Biochem. *9*:226.

83. De Meis, L. 1968. J. Biol. Chem. *243*:1174.

84. Deutscher, M. P., P. Chambon, and A. Kornberg. 1968. J. Biol. Chem. *243*:5117.

85. Dion, A., and S. S. Cohen. 1971. Fed. Proc. [in press].

86. Dion, A. S., and E. J. Herbst. 1967. Proc. Nat. Acad. Sci. *58*:2367.

87. Dion, A., and E. Herbst. 1970. Ann. N. Y. Acad. Sci. *171*:723.

88. Dobell, C. 1932. *Antony van Leeuwenhoek and His "Little Animals."* Republished in 1960 by Dover Publications, Inc., New York.

89. Donini, P. 1970. J. Bact. *102*:616.

89a. Driedger, A. A. 1970. Canad. J. Microbiol. *16*:881.

90. Dubin, D. T. 1959. Biochem. Biophys. Res. Commun. *1*:262.

91. Dubin, D. T. 1960. J. Biol. Chem. *235*:783.

92. Dubin, D. T., and S. M. Rosenthal. 1960. J. Biol. Chem. *235*:776.

93. Dudley, H. W., O. Rosenheim, and W. W. Starling. 1926. Biochem. J. *20*:1082.

94. Dudley, H. W., O. Rosenheim, and W. W. Starling. 1927. Biochem. J. *21*:97.

95. Dykstra, W. G., Jr., and E. J. Herbst. 1965. Science *149*:428.

96. Edlin, G., and P. Broda. 1968. Bact. Rev. *32*:206.

97. Eilon, G., and U. Bachrach. 1969. Biochim. Biophys. Acta *179*:464.

98. Elliott, B., and I. A. Michaelson. 1967. Anal. Biochem. *19*:184.

99. Elliott, B. and I. A. Michaelson. 1969. Proc. Soc. Exp. Biol. Med. *131*:105.

100. Ennis, H. L., and M. Lubin. 1961. Biochim. Biophys. Acta *50*:399.

101. Epstein, W., and M. Davies. 1970. J. Bact. *101*:836.

102. Epstein, W., and S. G. Schultz. 1965. J. Gen. Physiol. *49*:221.

103. Erdmann, V. A., G. A. Thomas, J. W. Norton, and E. J. Herbst. 1968. Biochim. Biophys. Acta. *157*:43.

104. Ezekiel, D. H., and H. Brockman. 1968. J. Mol. Biol. *31*:541.

105. Fausto, N. 1969. Biochim. Biophys. Acta *190*:193.

106. Fausto, N. 1970. Cancer Res. *30*:1947.

107. Felsenfeld, G., and S. L. Huang. 1961. Biochim. Biophys. Acta *51*:19.

108. Ferroluzzi-Ames, G., and B. N. Ames. 1965. Biochem. Biophys. Res. Commun. *18*:639.

109. Fiil, N., and J. D. Friesen. 1968. J. Bact. *95*:729.

110. Forbes, M., N. A. Kuck, and E. A. Peets. 1965. J. Bact. *89*:1299.

111. Forbes, M., E. A. Peets, and N. A. Kuck. 1966. Ann. N. Y. Acad. Sci. *135*:726.

112. Fox, C. F., R. I. Gumport, and S. B. Weiss. 1965. J. Biol. Chem. *240*:2101.

113. Fox, C. F., W. S. Robinson, R. Haselkorn, and S. B. Wess. 1964. J. Biol. Chem. *239*:186.

114. Fox, C. F., and S. B. Weiss. 1964. J. Biol. Chem. *239*:175.

115. Fraser, D., and H. R. Mahler. 1958. J. Am. Chem. Soc. *80*:6456.

116. Fraser, D., and H. R. Mahler. 1961. Biochim. Biophys. Acta *53*:199.

117. Fresco, J. 1970. Paper presented to Symp. N. Y. Acad. Sci. April, 1970.

118. Fresco, J. R., A. Adams, R. Ascione, D. Henley, and T. Lindahl. 1966. Cold Spring Harbor Symp. Quant. Biol. *31*:527.

119. Fresco, J. R., R. D. Blake, and R. Langridge. 1968. Nature *220*:1285.

120. Friedman, A. H., and L. D. Rodichok. 1970. Fed. Proc. *29*:617.

121. Friendman, S. M., and I. B. Weinstein. 1964. Proc. Nat. Acad. Sci. *52*:988.

122. Fuchs, E., R. L. Millette, W. Zillig, and G. Walter. 1967. European J. Biochem. *3*:183.

123. Gabbay, E. J., R. Glasser, and B. Gaffney. 1970. Ann. N. Y. Acad. Sci. *171*:810.

124. Gabbay, E. J., and J. Mitschele. 1969. Biochem. Biophys. Res. Commun. *34*:53.

125. Gabbay, E. J., and R. R. Shimshak. 1968. Biopolymers *6*:255.

126. Gale, E. F. 1940. Biochem. J. *34*:392.

127. Garen, A. 1961. Virology *14*:151.

128. Gartland, W. J., and N. Sueoka. 1966. Proc. Nat. Acad. Sci. *55*:948.

129. Gefter, M., R. Hausmann, M. Gold, and J. Hurwitz. 1966. J. Biol. Chem. *241*:1995.

130. Gershey, E. L., G. Vidali, and V. G. Allfrey. 1968. J. Biol. Chem. *243*:5018.

131. Giri, C. P., and S. V. Bhide. Indian J. Exp. Biol. *7*:95.

132. Glasser, R., and E. J. Gabbay. 1968. Biopolymers *6*:243.

133. Goldstein, J. 1964. Exp. Cell Res. *37*:494.

134. Goldstein, J. 1966. Biochim. Biophys. Acta *123*:620.

135. Gorkin, V. Z., and L. V. Tatyanenko. 1967. Biochem. Biophys. Res. Commun. *27*:613.

136. Greene, R. C., 1957. J. Am. Chem. Soc. *79*:3929.

137. Groman, N. B., and G. Suzuki. 1966. J. Bact. *92*:1735.

138. Grossowicz, N., and M. Ariel. 1963. J. Bact. *85*:293.

139. Gschwender, H. H., and P. H. Hofschneider. 1969. Biochem. Biophys. Acta *190*:454.

140. Guggenheim, M. 1951. *Die biogenen Amine*. 4th ed. S. Karger, Basel.

141. Guirard, B. M., and E. E. Snell. 1964. J. Bact. *88*:72.

142. Gumport, R. F., and S. B. Weiss. 1969. Biochemistry *8*:3618.

143. Guttman, B. S., and L. Begley. 1969. Virology *36*:687.

144. Halevy, S., Z. Fuchs, and J. Mager. 1962. Bull. Res. Counc. Israel, Sect. A. *11*:52.

145. Ham, R. G. 1964. Biochem. Biophys. Res. Comm. *14*:34.

146. Hammond, J. E., and E. J. Herbst. 1968. Anal. Biochem. *22*:474.

147. Hampel, A., and R. Bock. 1970. Biochemistry *9*:1873.

148. Harold, F. M. 1964. J. Bact. *88*:1416.

149. Heptinstall, S., A. R. Archibald, and J. Baddiley. 1970. Nature *225*:519.

150. Herbst, E. J., and A. S. Dion. 1970. Fed. Proc. *29*:1563.

151. Herbst, E. J., and B. P. Doctor. 1959. J. Biol. Chem. *234*:1497.

152. Herbst, E. J., E. B. Glinos, and L. Amundsen. 1955. J. Biol. Chem. *214*:175.

153. Herbst, E. J., and E. E. Snell. 1948. J. Biol. Chem. *176*:989.

154. Herbst, E. J., R. H. Weaver, and D. L. Keister. 1958. Arch. Biochem. Biophys. *75*:171.

155. Heredia, C. F., and H. O. Halvorson. 1966. Biochemistry *5*:946.

156. Hershey, A. D. 1957. Virology *4*:237.

157. Hershko, A., S. Amoz, and J. Mager. 1961. Biochem. Biophys. Res. Commun. *5*:46.

158. Herzfeld, A., and O. Greengard. 1969. J. Biol. Chem. *244*:4894.

159. Hettinger, T. P., and L. C. Craig. 1970. Biochemistry *9*:1224.

160. Hiebert, E., and J. B. Bancroft. 1960. Virology *39*:296.

161. Higgins, M. L., M. C. Tillman, J. P. Rupp, and F. R. Leach. 1969. J. Cell. Physiol. *74*:149.

162. Higuchi, S., and M. Tsuboi. 1966. Bull. Chem. Soc. Japan *39*:1886.

163. Hirashima, A., K. Asano, and A. Tsugita. 1967. Biochim. Biophys. Acta *134*:165.

164. Hirsch, J. G. 1953. J. Exp. Med. *97*:327.

165. Hirsch, J. G. 1953. J. Exp. Med. *97*:345.

166. Hirsch, J. G., and R. J. Dubos. 1952. J. Exp. Med. *95*:191.

167. Hirschberg, E., I. B. Weinstein, N. Gersten, E. Marner, T. Finkelstein, and R. Carchman. 1968. Cancer Res. *28*:601.

168. Hirschfield, I. N., R. De Deken, P. C. Horn, D. A. Hopwood, and W. K. Maas. 1968. J. Mol. Biol. *35*:83.

169. Hirschfield, I. N., H. J. Rosenfeld, Z. Leifer, and W. K. Maas. 1970. J. Bact. *101*:725.

170. Hirschman, S. Z., M. Leng, and G. Felsenfeld. 1967. Biopolymers *5*:227.

171. Hochstein, L. I., and B. P. Dalton. 1968. Biochim. Biophys. Acta *167*:638.

172. Hodgins, D. S., and Abeles, R. H. 1969. Arch. Biochem. Biophys. *130*:274.

173. Horáček, P., and I. J. Cernohorsky. 1968. Biochem. Biophys. Res. Commun. *32*:956.

174. Hurlbert, R. B., E. G. Miller, and C. L. Vaughan. 1969. *Advances in Enzyme Regulation 7*, p. 219, Pergamon Press, Oxford and New York.

175. Hurwitz, C., and C. L. Rosano. 1967. J. Biol. Chem. *242*:3719.

176. Igarashi, K., and Y. Takeda. 1970. Biochim. Biophys. Acta *213*:240.

177. Iliev, V., R. E. Nichols, and C. C. Pfeiffer. 1968. Anal Biochem. *25*:1.

178. Inners, L. D., and G. Felsenfeld. 1970. J. Mol. Biol. *50*:373.

179. Inouye, M., and A. B. Pardee. 1970. J. Bact. *101*:770.

180. Ishida, T., and N. Sueoka. 1967. Proc. Nat. Acad. Sci. *58*:1080.

181. Ishida, T., and N. Sueoka. 1968. J. Mol. Biol. *37*:313.

182. Ishida, T., and N. Sueoka. 1968. J. Biol. Chem. *243*:5329.

183. Israel, M., J. S. Rosenfield, and E. J. Modest. 1964. J. Med. Chem. *7*:710.

184. Jackson, E. L. 1956. J. Org. Chem. *21*:1374.

185. Jamieson, G. A. 1963. J. Org. Chem. *28*:2397.

186. Jamieson, G. A. 1968. *In* W. Zorbach and S. Tipson [eds.] *Synthetic Procedures in Nucleic Acid Chemistry 1,* p. 176, Interscience, New York.

187. Jänne, J. 1967. Acta Physiol. Scand. (suppl.) *300*:7.

188. Jänne, J., and A. Raina. 1966. Acta Chem. Scand. *20*:1174.

188a. Jänne, J., and H. G. Williams-Ashman. 1970. Biochem. J. *119*:595.

188b. Jänne, J., and H. G. Williams-Ashman [in press].

189. Jänne, J., and A. Raina. 1968. Acta Chem. Scand. *22*:1349

190. Jänne, J., and A. Raina. 1969. Biochim. Biophys. Acta *174*:769.

191. Jänne, J., A. Raina, and M. Siimes. 1964. Acta Physiol. Scand. *62*:352.

192. Jänne, J., A. Raina, and M. Siimes. 1968. Biochim. Biophys. Acta *166*:419.

193. Jellinck, P. H., and G. Perry. 1967. Biochim. Biophys. Acta *137*:367.

194. Johnson, H. G., and M. K. Bach. 1965. Nature *208*:408.

195. Johnson, H. G., and M. K. Bach. 1966. Proc. Nat. Acad. Sci. *55*:1453.

196. Johnson, H. G., and M. K. Bach. 1968. Arch. Biochem. Biophys. *128*: 113.

197. Johnson, M. W., and R. Markham. 1962. Virology *17*:276.

198. Kaiser, D., H. Tabor, and C. W. Tabor. 1963. J. Mol. Biol. *6*:141.

199. Kakimoto, Y., T. Nakajima, A. Kumon, Y. Matsuoka, N. Imaoka, I. Sano, and A. Kanagawa. 1969. J. Biol. Chem. *244*:6003.

200. Kedracki, R., and W. Szer. 1967. Acta Biochim. Polonica *14*:163.

201. Kekomäki, M. P., N. C. R. Räihä, and H. Bickel. 1969. Clin. Chim. Acta *23*:203.

202. Kersaint, G. 1958. Bull. Soc. Chim. France 1603.

202a. Khawaja, J. A., and A. Raina. 1970. Biochem. Biophys. Res. Commun. *41*:512.

203. Khawaja, J. A., and L. Stevens. 1967. Proc. Biochem. Soc. *104*:13.

203a. Khesin, R. B., S. Z. Mindlin, Zh. M. Gorlenko, and T. S. Ilyina. 1968. Molec. Gen. Genetics *103*:194.

204. Kihara, H., and E. E. Snell. 1957. Proc. Nat. Acad. Sci. *43*:867.

205. Kim, K. H. 1963. J. Bact. *86*:320.

206. Kim, K. 1966. J. Bact. *91*:193.

207. King, H. 1959. *Sigmund Otto Rosenheim: Biographical Memoirs of Fellows of the Royal Socety*. Royal Society, London.

208. Kostyo, J. L. 1966. Biochem. Biophys. Res. Commun. *23*:150.

209. Krakow, J. S. 1963. Biochim. Biophys. Acta *72*:566.

210. Krakow, J. S. 1966. J. Biol. Chem. *241*:1830.

211. Krakow, J. S., and S. Ochoa. 1963. Biochem. Z. *338*:796.

212. Kremzner, L. T. 1966. Anal. Biochem. *15*:270.

213. Kremzner, L. T. 1970. Fed. Proc. *29*:1583.

214. Kremzner, L. T., and L. J. Cote. 1968. Fed. Proc. *27*:839.

215. Kurland, C. G., and O. Maaløe. 1962. J. Mol. Biol. *4*:193.

216. Kwiatkowski, Z., and H. Heleszko. 1969. Acta Microbiol. Polonica *1*:101.

217. Labanauskas, M., P. G. Connors, J. D. Young, R. M. Bock, J. W. Anderegg, and W. W. Beeman. 1970. Science *166*:1530.

218. Lampson, G. P., A. A. Tytell, A. K. Field, M. M. Nemes, and M. F. Hilleman. 1969. Proc. Soc. Exp. Biol. and Med. *132*:212.

219. Lanning, M. C., and S. S. Cohen. 1954. J. Biol. Chem. *207*:193.

220. Lazzarini, R. A., and E. Santangelo. 1968. J. Bact. *95*:1212.

221. Leboy, P. S. 1970. Biochemistry *9*:1577.

222. Lee-Huang, S., and R. C. Warner. 1969. J. Biol. Chem. *244*:3793.

223. Leeuwenhoek, A. 1678. Phil. Trans. Roy. Soc. *12*:1040.

224. Levitt, M. 1969. Nature *224*:759.

225. Lindahl, T., and A. Adams. 1966. Science *152*:512.

226. Lindahl, T., A. Adams, and J. R. Fresco. 1966. Proc. Nat. Acad. Sci. *55*:941.

227. Lindahl, T., A. Adams, and J. R. Fresco. 1967. J. Biol. Chem. *242*:3129.

228. Lindahl, T., A. Adams, M. Geroch, and J. R. Fresco. 1967. Proc. Nat. Acad. Sci. *57*:178.

229. Lindahl, T., D. D. Henley, and J. R. Fresco. 1965. J. Am. Chem. Soc. *87*:4961.

230. Liquori, A. M., F. Ascoli, and M. De Santis Savino. 1967. J. Mol. Biol. *24*:123.

231. Liquori, A. M., L. Constantino, V. Crescenzi, V. Elia, E. Giglio, R. Puliti, M. De Santis Savino, and V. Vitagliano. 1967. J. Mol. Biol. *24*:113.

232. Loeb, M. R., and S. S. Cohen. 1959. J. Biol. Chem. *234*:364.

233. Lubin, M. 1964. Fed. Proc. *23*:994.

234. Lubochinsky, B., J. Meury, and J. Stolkowski. 1964. C. R. Acad. Sci. Paris *258*:5106.

235. Maas, W. K., Z. Leifer, and J. Poindexter. 1970. Ann. N. Y. Acad. Sci. *171*:957.

236. Mager, J. 1955. Nature *176*:933.

237. Mager, J. 1959. Biochim. Biophys. Acta *36*:429.

238. Mager, J., M. Benedict, and M. Artman. 1962. Biochim. Biophys. Acta *62*:202.

239. Mahler, H. R., and M. B. Baylor. 1967. Proc. Nat. Acad. Sci. *58*:256.

240. Mahler, H. R., and G. Dutton. 1964. J. Mol. Biol. *10*:157.

241. Mahler, H. R., R. Goutarel, Q. Khuong-Huu, and M. T. Ho. 1966. Biochemistry *5*:2177.

242. Mahler, H. R., G. Green, R. Goutard, and Q. Khuong-Huu. 1968. Biochemistry *7*:1568.

243. Mahler, H. R., and B. D. Mehrotra. 1963. Biochim. Biophys. Acta *68*:211.

244. Mahler, H. R., B. D. Mehrottra, and C. W. Sharp. 1961. Biochem. Biophys. Res. Commun. *4*:79.

245. Maitra, U., and F. Barash. 1969. Fed. Proc. *28*:659.

246. Maitra, U., and J. Hurwitz. 1967. J. Biol. Chem. *242*:4897.

247. Mandel, M. 1962. J. Mol. Biol. *5*:435.

248. Marsh, J. B., D. L. Drabkin, and L. Poppe. 1970. Metabolism *19*:392.

249. Marshall, M., R. S. Metzenberg, and P. P. Cohen. 1961. J. Biol. Chem. *236*:2229.

250. Martin, R. G., and B. N. Ames. 1962. Proc. Nat. Acad. Sci. *48*:2171.

251. Martin, W. H., Jr., M. J. Pelczar, Jr., and P. A. Hansen. 1952. Science *116*:483.

252. Matsuo, K., and M. Tsuboi. 1966. Bull. Chem. Soc. Japan *39*:347.

253. Matzura, H., and P. Broda. 1968. J. Bact. *96*:1877.

254. Meeuse, B. J. D. 1966. Scientific Am. *215*:80.

255. Mehrotra, B. D., and H. R. Mahler. 1964. Biochim. Biophys. Acta *91*:78.

256. Michaels, R., and K. H. Kim. 1966. Biochim. Biophys. Acta *115*:59.

257. Michaels, R., and T. T. Tchen. 1968. J. Bact. *95*:1966.

257. Michaels, R., J. Tobari, and T. T. Tchen. 1967. Fed. Proc. *26*:870.

259. Michaelson, I. A., and P. Z. Coffman. 1963. Biochem. Pharmacol. *16*: 2085.

260. Millette, R. L. 1969. Fed. Proc. *28*:659.

261. Mills, J., and D. T. Dubin. 1966. Mol. Pharm. *2*:311.

262. Mitra, S., and P. Kaesberg. 1963. Biochem. Biophys. Res. Commun. *11*:146.

263. Miyazawa, F., V. C. Dick, and T. Tamaoki. 1968. Biochim. Biophys. Acta *115*:193.

264. Moller, M. L., and K. Kim. 1965. Biochem. Biophys. Res. Commun. *20*:46.

265. Morris, D. R., and C. M. Jorstad. 1970. J. Bact. *101*:731.

266. Morris, D. R., and K. L. Koffron. 1967. J. Bact. *94*:1516.

267. Morris, D. R., and K. L. Koffron. 1969. J. Biol. Chem. *244*:6094.

268. Morris, D. R., and A. B. Pardee. 1965. Biochem. Biophys. Res. Commun. *20*:697.

269. Morris, D. R., and A. B. Pardee. 1966. J. Biol. Chem. *241*:3129.

270. Morris, D. W., and J. A. De Moss. 1965. J. Bact. *90*:1624.

271. Moruzzi, G., B. Barbiroli, and C. M. Caldarera. 1968. Biochem. J. *107*:609.

272. Mudd, S. H., and G. L. Cantoni. 1958. J. Biol. Chem. *231*:481.

273. Muench, K. H. 1966. Cold Spring Harbor Symp. Quant. Biol. *31*:539.

274. Muench, K. H. 1969. Biochemistry *8*:4872.

275. Muench, K. H. 1969. Biochemistry *8*:4880.

276. Muench, K. H., and P. A. Safille. 1968. Biochemistry *7*:2799.

277. Nakamura, K., and T. Tamaoki. 1968. Biochim. Biophys. Acta *161*:368.

278. Nakajima, T., J. F. Zack, and F. Wolfgram. 1969. Biochim. Biophys. Acta *184*:651.

279. Nathans, D., and F. Lipmann. 1961. Proc. Nat. Acad. Sci. *47*:497.

280. Nazar, R. N., and J. T. Wong. 1969. J. Bact. *100*:956.

281. Nierlich, D. P. 1967. Science *158*:1186.

282. Nierlich, D. P. 1968. Proc. Nat. Acad. Sci. *60*:1345.

283. O'Brien, R. L., J. L. Allison, and F. E. Hahn. 1966. Biochim. Biophys. Acta *129*:622.

284. Ochoa, M., Jr., and I. B. Weinstein. 1965. Biochim. Biophys. Acta *95*:176.

285. Oki, T., H. Kawasaki, K. Ogata, H. Yamada, I. Tomida, T. Morino, and H. Fukami. 1968. Agr. Biol. Chem. *32*:1349.

286. Okuyama, T., and Y. Kobayashi. 1961. Arch. Biochem. Biophys. *95*:242.

287. Olenick, J. G., and F. E. Hahn. 1966. Biochim. Biophys. Acta *129*:429.

288. Padmanabhan, R., and K. Kim. 1965. Biochem. Biophys. Res. Commun. *19*:1.

289. Passero, F., E. J. Gabbay, B. Gaffney, and T. Kurucsev. 1970. Macromolecules *3*:158.

290. Pearce, L. A., and S. M. Schanberg. 1969. Science *166*:1301.

291. Pegg, A. E., and H. G. Williams-Ashman. 1968. Biochem. Biophys. Res. Commun. *30*:76.

292. Pegg, A. E., and H. G. Williams-Ashman. 1968. Biochem. J. *108*:533.

293. Pegg, A. E., and H. G. Williams-Ashman. 1968. J. Biol. Chem. *244*:682.

294. Pegg, A. E., and H. G. Williams-Ashman. 1970. Arch. Biochem. Biophys. *137*:156.

295. Perry, T. L., S. Hansen, and L. MacDougall. 1967. J. Neurochem. *14*:775.

296. Petersen, E. E., H. Kröger, and U. Hagen. 1968. Biochim. Biophys. Acta *161*:325.

297. Pett, D. M., and H. S. Ginsberg. 1968. Fed. Proc. *27*:615.

298. Philipps, G. R. 1969. Nature *223*:374.

299. Powell, G., K. V. Rajagopalan, and P. Handler. J. Biol. Chem. *244*:4793.

300. Quigley, J. W., and S. S. Cohen. 1969. J. Biol. Chem. *244*:2450.

301. Räihä, N. C. R., and M. P. Kekomäki. 1968. Biochem. J. *108*:521.

302. Raina, A. 1962. Acta Chem. Scand. *16*:2463.

303. Raina, A. 1963. Acta Physiol. Scand. *60*(Suppl. 218):1.

304. Raina, A. 1969. Symp. Biogenic Amines, Woods Hole, Mass. [In press].

305. Raina, A., and S. S. Cohen. 1966. Proc. Nat. Acad. Sci. *55*:1587.

305a. Raina, A., and P. Hannonen. Acta Chem. Scand. [In press].

306. Raina, A., and J. Jänne. 1968. Acta Chem. Scand. *22*:2375.

307. Raina, A., and J. Jänne. 1970. Fed. Proc. *29*:1568.

308. Raina, A., J. Jänne, and M. Siimes. 1966. Biochim. Biophys. Acta *123*:197.

309. Raina, A., M. Jansen, and S. S. Cohen. 1967. J. Bact. *94*:1684.

310. Raina, A., M. Siimes, and J. Jänne. 1968. Scand. J. Clin. Lab. Invest. *21*:10.

311. Raina, A., and T. Telaranta. 1967. Biochim. Biophys. Acta *138*:200.

312. Razin, S. 1964. J. Gen. Microbiol. *36*:451.

313. Razin, S., A. Cohen, and R. Rozansky. 1958. Proc. Soc. Exp. Biol. Med. *99*:459.

314. Razin, S., and R. Rozansky. 1959. Arch. Biochem. Biophys. *81*:36.

315. Reichard, P. 1962. J. Biol. Chem. *237*:2513.

316. Reiter, H. 1963. Virology *21*:636.

317. Rhodes, J. B., and H. G. Williams-Ashman. 1964. Medicina Experimentalis *10*:21.

318. Richards, F. J., and R. G. Coleman. 1952. Nature *170*:460.

319. Riley, W. D., and E. E. Snell. 1968. Biochemistry *7*:3520.

320. Rinaldini, L. M., and W. K. Maas. 1969. J. Cell Biol. *39*:111(a).

321. Robin, Y., C. Audit, and M. Landon. 1967. Comp. Biochem. Physiol. *22*:787.

322. Robin, Y., and J. Roche. 1965. Comp. Biochem. Physiol. *14*:453.

323. Robin, Y., and N. van Thoai. 1961. C. R. Acad. Sci. Paris *252*:1224.

324. Roche, J., C. Audit, and Y. Robin. 1965. C. A. Acad. Sci. Paris *260*: 7023.

325. Rodwell, A. W. 1965. J. Gen. Microbiol. *40*:227.

326. Rogers, G. T., T. L. V. Ulbricht, and W. Szer. 1967. Biochem. Biophys. Res. Commun. *27*:372.

327. Ron, E. Z., and B. D. Davis. 1966. J. Mol. Biol. *21*:13.

328. Rosano, C. L., and C. Hurwitz. 1969. Biochem. Biophys. Res. Commun. *37*:677.

329. Rosenheim, Otto. 1924. Biochem. J. *18*:1253.

330. Rosenthal, S. M., and D. T. Dubin. 1962. J. Bact. *84*:859.

331. Rosenthal, S. M., E. R. Fisher, and E. F. Stohlman, 1952. Proc. Soc. Exp. Biol. Med. *80*:432.

332. Rosenthal, S. M., and C. W. Tabor. 1956. J. Pharmacol. Exp. Therap. *116*:131.

333. Rubenstein, K. E., and S. S. Cohen. 1971. J. Bact. [In press].

334. Russell, D. 1970. Fed. Proc. *29*:669.

335. Russell, D. H. 1970. Proc. Am. Assoc. Cancer Res. *11*:69

335a. Russell, D. H., V. J. Medina, and S. H. Snyder. 1970. J. Biol. Chem. *245*:6732.

336. Russell, D., and S. Snyder. 1968. Proc. Nat. Acad. Sci. *66*:1420.

337. Russell, D., and S. Snyder. 1969. Mol. Pharm. *5*:253.

338. Russell, D., and S. Snyder. 1969. Endocrinology *84*:223.

339. Russell, D. H., S. H. Snyder, and V. J. Medina. 1969. Life Science *8*:1247.

340. Russell, D. H., S. H. Snyder, and V. J. Medina. 1970. Endocrinology *86*:1414.

341. St. Amand, G. A., N. G. Anderson, and M. E. Gaulden. 1960. Exp. Cell. Res. *20*:71.

342. Sakiyama, S., S. Usui, and Y. Miura. 1969. *Advances in Enzyme Regulation 7*:207. Pergamon Press, Oxford and New York.

343. Sakurada, T., and H. Matsumura. 1964. J. Biochem. (Japan) *56*:208.

344. Salganicoff, L. 1968. Fed. Proc. *27*:527.

345. Sanwal, B. D. 1970. J. Biol. Chem. *245*:1626.

346. Sanwal, B. D., P. Maeba, and R. A. Cook. 1966. J. Biol. Chem. *241*: 5177.

347. Savino, M., and F. Ascoli. 1969. Biochim. Biophys. Acta *195*:569.

348. Schlenk, F., and J. L. Dainko. 1966. Arch. Biochem. Biophys. *113*:127.

349. Schrock, T. R., N. J. Oakman, and N. L. R. Bucher. 1970. Biochim. Biophys. Acta *204*:564.

350. Schroeder, H. R., and M. F. Mallette. 1970. Fed. Proc. *29*:925.

351. Scott, D. B. M., and S. S. Cohen. 1953. Biochem. J. *55*:23.

352. Seiler, N., G. Werner, and H. A. Fischer. 1969. Z. Physiol. Chem. *350*:676.

353. Seiler, N., and M. Wiechmann. 1967. Z. Physiol. Chem. *348*:1285.

354. Sevag, M. G., and W. T. Drabble. 1962. Biochem. Biophys. Res. Commun. *8*:446.

355. Shalitin, C., and S. Sarid. 1967. J. Virology *1*:559.

356. Shapiro, J. T., B. S. Stannard, and G. Felsenfeld. 1969. Biochemistry *8*:3233.

357. Shapiro, S. K., and A. N. Mather. 1958. J. Biol. Chem. *233*:631.

358. Shilo, M. 1967. Bact. Rev. *31*:180.

359. Shilo, M., and M. Shilo. 1962. J. Gen. Microbiol. *29*:645.

360. Shimizu, H., Y. Kakimoto, and I. Sano. 1964. J. Pharm. Exp. Therap. *143*:199.

361. Shimizu, H., Y. Kakimoto, and I. Sano. 1965. Arch. Biochem. Biophys. *110*:368.

362. Shimizu, H., Y. Kakimoto, and I. Sano. 1965. Nature *207*:1196.

363. Siekevitz, P., and G. E. Palade. 1962. J. Cell Biol. *13*:217.

364. Siimes, M. 1967. Acta Physiol. Scand. (suppl.) *298*:7.

365. Sillman, N., M. Artman, and H. Engelberg. 1965. Biochem. Biophys. Acta *103*:231.

366. Silver, S., and D. Clark. 1970. Bact. Proc. *130*:47.

367. Silver, S., and M. L. Kralovic. 1969. Mol. Pharm. *5*:300.

368. Silver, S., and E. Levine. 1968. J. Bact. *96*:338.

369. Silver, S., and E. Levine. 1968. Biochem. Biophys. Res. Commun. *31*:743.

370. Silverman, M., and E. A. Evans, Jr. 1944. J. Biol. Chem. *154*:521.

371. Simon, E. J., S. S. Cohen, and A. Raina. 1966. Biochem. Biophys. Res. Commun. *24*:482.

371a. Simon, E. J., L. Schapira, and N. Wurster. 1970. Mol. Pharmacol. *6*:577.

372. Simon, E. J., and D. Van Praag. 1964. Proc. Nat. Acad. Sci. *51*:1151.

373. Simpkins, H., and E. G. Richards. 1967. J. Mol. Biol. *29*:349.

374. Sinclair, C. 1967. Nature *213*:214.

375. Smeaton, W. A. 1962. *Fourcroy: Chemist and Revolutionary*. W. Heffer and Sons, Ltd., Cambridge.

375a. Smeaton, W. A. 1963. Platinum Metals Rev. *7*:106.

376. Smith, B. N., and B. J. D. Meeuse. 1966. Plant Physiol. *41*:343.

377. Smith, R. C., and W. D. Salmon. 1965. J. Bact. *89*:687.

378. Smith, T. A. 1963. Phytochemistry *2*:241.

379. Smith, T. A. 1965. Phytochemistry *4*:599.

380. Smith, T. A. 1969. Phytochemistry *8*:2111.

381. Smith, T. A. 1970. Phytochemistry *9*:1479.

382. Smith, T. A., and F. J. Richards. 1962. Biochem. J. *84*:292.

383. Smith, T. A., and C. Sinclair. 1967. Ann Bot., N. S. *31*:103.

384. Smythies, J. R., and F. Antun. 1969. Nature *223*:1061.

385. Sneath, P. H. A. 1955. Nature *175*:818.

386. Snell, E. E. 1944. J. Biol. Chem. *152*:475.

387. Snyder, S. H., and E. D. Hendley. 1968. J. Pharm. Exp. Therap. *163*:386.

388. So, A. G., E. W. Davie, R. Epstein, and A. Tissières. 1967. Proc. Nat. Acad. Sci. *58*:1739.

389. Soda, K., and M. Moriguchi. 1969. Biochem. Biophys. Res. Commun. *34*:34.

390. Sokawa, Y., E. Nakao-Sato, and Y. Kaziro. 1970. Biochim. Biophys. Acta *199*:256.

391. Sprince, H., and D. W. Wooley. 1944. J. Exp. Med. *80*:213.

392. Stastny, M., and S. Cohen. 1970. Biochim. Biophys. Acta *204*:578.

393. Steele, W. J., N. Okamura, and H. Busch. 1965. J. Biol. Chem. *240*: 1742.

394. Stent, G. S., and S. Brenner. 1961. Proc. Nat. Acad. Sci. *47*:2005.

395. Stern, J. L., H. D. Barner, and S. S. Cohen. 1966. J. Mol. Biol. *17*:188.

396. Stern, J. L., M. Sekiguchi, H. D. Barner, and S. S. Cohen. 1964. J. Mol. Biol. *8*:629.

397. Stevens, L. 1966. Proc. Biochem. Soc. *18*:11p.

398. Stevens, L. 1967. Biochem. J. *103*:811.

399. Stevens, L. 1970. Biol. Rev. *45*:1.

400. Stevens, L., and M. R. Morrison. 1968. Biochem. J. *108*:633.

400a. Stirpe, F., and F. Novello. 1970. Eur. J. Biochem. *15*:505.

401. Suelter, C. H. 1970. Science *168*:789.

402. Sussman, A. J., and C. Gilvarg. 1969. J. Biol. Chem. *244*:6304.

402a. Suwalsky, M., W. Traub, U. Shmual, and J. A. Subirana. 1969. J. Mol. Biol. *42*:363.

403. Szer, W. 1966. J. Mol. Biol. *16*:585.

404. Szer, W. 1966. Biochem. Biophys. Res. Commun. *22*:559.

405. Szer, W. 1966. Acta Biochem. Polonica *13*:251.

406. Tabor, C. W. 1962. J. Bact. *83*:1101.

407. Tabor, C. W. 1968. Biochem. Biophys. Res. Commun. *30*:339.

408. Tabor, C. W. 1962. *In* S. P. Colowick and N. O. Kaplan [eds.] *Methods in Enzymology*. Academy Press, New York.

409. Tabor, C. W., and L. G. Dobbs. 1970. J. Biol. Chem. *245*:2086.

410. Tabor, C. W., and P. D. Kellogg. 1967. J. Biol. Chem. *242*:1044.

411. Tabor, C. W., and P. D. Kellogg. 1968. Fed. Proc. *27*:639.

412. Tabor, C. W., and H. Tabor. 1966. J. Biol. Chem. *241*:3714.

412a. Tabor, C. W., and H. Tabor. 970. Biochem. Biophys. Res. Commun. *41*:232.

413. Tabor, C. W., H. Tabor, and U. Bachrach. 1964. J. Biol. Chem. *239*:2194.

414. Tabor, C. W., H. Tabor, and L. De Meis. 1966. Fed. Proc. *25*:709.

415. Tabor, C. W., H. Tabor, and L. G. Dobbs. 1967. Fed. Proc. *26*:725.

416. Tabor, H. 1960. Biochem. Biophys. Res. Commun. *3*:382.

417. Tabor, H. 1961. Biochem. Biophys. Res. Commun. *4*:228.

418. Tabor, H. 1962. Biochemistry *1*:496.

419. Tabor, H., S. M. Rosenthal, and C. W. Tabor. 1957. J. Amer. Chem. Soc. *79*:2978.

420. Tabor, H., S. M. Rosenthal, and C. W. Tabor. 1958. J. Biol. Chem. *233*:907.

421. Tabor, H., and C. W. Tabor. 1964. Pharm. Rev. *16*:245.

422. Tabor, H., and C. W. Tabor. 1966. Fed. Proc. *25*:879.

423. Tabor, H., and C. W. Tabor. 1969. J. Biol. Chem. *244*:2286.

424. Tabor, H., and C. W. Tabor. 1969. J. Biol. Chem. *244*:6383.

425. Tabor, H., C. W. Tabor, and S. M. Rosenthal. 1961. Ann. Rev. Biochem. *30*:579.

426. Takeda, Y. 1969. J. Biochem. (Japan) *66*:345.

427. Takeda, Y. 1969. Biochim. Biophys. Acta *179*:232.

428. Takeda, Y. 1969. Biochim. Biophys. Acta *182*:258.

429. Takeda, Y., and K. Igarashi. 1969. Biochim. Biophys. Res. Commun. *37*:917.

430. Takeda, Y., and K. Igarashi. 1970. Biochim. Biophys. Acta *204*:406.

431. Tal, M. 1969. Biochim. Biophys. Acta *195*:76.

432. Tanner, M. J. A. 1967. Biochemistry *6*:2686.

433. Tempest, D. W., J. W. Dicks, and J. R. Hunter. 1966. J. Gen. Microbiol. *45*:135.

434. Tissières, A., S. Bourgeois, and F. Gros. 1963. J. Mol. Biol. *7*:100.

435. Tomita, T. 1962. Hoppe-Seyl. Z. *328*:265.

436. Traub, A., J. Mager, and N. Grossowicz. 1955. J. Bact. *70*:60.

437. Tropp, B. E., L. C. Meade, and P. J. Thomas. 1970. J. Biol. Chem. *245*:855.

438. Tsuboi, M. 1964. Bull. Chem. Soc. Japan *37*:1514.

439. Tucker, R. G. 1963. J. Gen. Microbiol. *32*:287.

440. Ulitzur, S., and M. Shilo. 1964. J. Gen. Microbiol. *36*:161.

441. Unemoto, T. 1963. Chem. Pharm. Bull. *11*:1255.

442. Unemoto, T., K. Ikeda, M. Hayashi, and K. Miyaki. 1963. Chem. Pharm. Bull. *11*:148.

443. Vauquelin, N. L. 1791. Annales de Chemie (première série) *9*:64.

444. Venis, M. A. 1969. Phytochemistry *8*:1193.

445. Wang, J. H., P. Humniski, and W. J. Black. 1968. Biochemistry *7*:2037.

446. Weaver, R. H., and E. J. Herbst. 1958. J. Biol. Chem. *231*:637.

447. Weaver, R. H., and E. J. Herbst. 1958. J. Biol. Chem. *231*:647.

448. Weinstein, I. B., R. Carchman, E. Marner, and E. Hirschberg. 1967. Biochim. Biophys. Acta *142*:440.

449. Weinstein, I. B., M. Ochoa, Jr., and S. M. Friedman. 1966. Biochemistry *5*:3332.

450. Weiss, R. L., and D. R. Morris. 1970. Biochim. Biophys. Acta *204*:502.

451. Weller, D. L., A. Raina, and D. B. Johnstone. 1968. Biochim. Biophys. Acta *157*:558.

452. Wheaton, T. A., and I. Stewart. 1965. Nature *206*:620.

453. Whitfield, J. F., R. H. Rixon, and T. Youdale. 1962. Exp. Cell Res. *27*:143.

454. Wickner, R. B., C. W. Tabor, and H. Tabor. 1970. J. Biol. Chem. *245*:2132.

455. Williams-Ashman, H. G. 1965. Investigative Urology *2*:605.

455a. Williams-Ashman, H. G., and D. H. Lockwood. 1970. Ann. N. Y. Acad. Sci. *171*:882.

456. Williams-Ashman, H. G., S. Liao, R. L. Hancock, L. Jurkowitz, and D. A. Silverman. 1964. Recent Progr. Hormone Res. *20*:247.

457. Williams-Ashman, H. G., A. E. Pegg, and D. H. Lockwood. 1969. *Advances in Enzyme Regulation 7*:291. Pergamon Press, Oxford and New York.

458. Winslow, R. M., and R. A. Lazzarini. 1969. J. Biol. Chem. *244*:3387.

459. Witkin, S. S., and E. Rosenberg. 1970. J. Bact. *103*:641.

460. Wu, W. H., and D. R. Morris. 1970. Bact. Proc. 68.

461. Yamada, H., O. Adachi, and K. Ogato. 1965. Agr. Biol. Chem. *29*:1148.

462. Yamada, H., T. Kimura, A. Tanaka, and K. Ogata. 1964. Agr. Biol. Chem. *28*:443.

463. Yoshida, D. 1969. Plant and Cell Physiol. *10*:393.

464. Yoshida, D., and T. Mitake. 1966. Plant and Cell Physiol. *7*:301.

465. Yoshikawa, H., and B. Marno. 1960. Biochim. Biophys. Acta *45*:270.

466. Young, J. D., R. M. Bock, S. Nishimura, H. Ishikura, Y. Yamada, U. L. Rajbhandary, M. Labanauskas, and R. G. Connors. 1969. Science *166*:1527.

467. Zappia, V., C. R. Zydek-Cwick, and F. Schlenk. 1969. J. Biol. Chem. *244*:4499.

468. Zillig, W., W. Krone, and M. Albers. 1959. Z. Physiol. Chem. *317*:131.

Index

A

D

G

H

O

P

T

U